Geopolitics

The term 'geopolitics' came into use at the end of the nineteenth century. Thinking globally was then formally connected by geopolitical reason to acting globally, but the actual practices of geopolitics began much earlier, when Europeans first encountered the rest of the world.

Geopolitics identifies and scrutinizes the central features of geopolitics from the sixteenth century to the present, paying close attention to its persisting conceptual underpinnings, novel turns and shifting impacts. The book focuses on four key concepts of the modern geopolitical imagination: visualizing the world as a whole; the definition of geographical areas as 'advanced' or 'primitive'; the notion of the state being the highest form of political organization; and the pursuit of primacy by competing states.

Exemplified by topical issues such as the reintegration of Hong Kong into China, the proposed expansion of powers of the EU at the expense of member states and the threatened break-up of states such as Canada, Spain, Russia and the UK by regionalist separatist movements, the book shows how questions of the organization of power combine with those of geographical definition and highlights the crucial geopolitical 'certainties' from as recently as ten years ago which have now either disappeared or are in question. *Geopolitics* provides an invaluable introduction to current, critical debates over 'geopolitics' and world politics.

John Agnew is Professor of Geography at the University of California, Los Angeles.

FRONTIERS

•

Edited by **Derek Gregory**, University of British Columbia, Canada and **Linda McDowell**, University of Cambridge, UK

An exciting new series of concise, accessible, yet thought-provoking books, focusing on the key ideas in contemporary geography and engaging students with new critical perspectives.

Geopolitics

re-visioning world politics

•

John Agnew

London and New York

First published 1998
by Routledge
11 New Fetter Lane, London EC4P 4EE

Simultaneously published in the USA and Canada
by Routledge
29 West 35th Street, New York, NY 10001

Typeset in Sabon by Keystroke, Jacaranda Lodge, Wolverhampton
Printed and bound in Great Britain by Butler and Tanner Ltd., Frome and London

British Library Cataloguing in Publication Data
A catalogue record for this book is available from the British Library

Library of Congress Cataloguing in Publication Data
Agnew, John A.
Geopolitics : re-visioning world politics / John Agnew.
p. cm. – (Frontiers of human geography ; 1)
Includes bibliographical references and index.
ISBN 0–415–14094–3 (cloth : alk. paper). – ISBN 0–415–14095–1
(paper : alk. paper)
1. Geopolitics. I. Title. II. Series.
JC319.A43 1998
320.1′2–dc21 97–35289
CIP

ISBN 0–415–14094–3 (hbk)
ISBN 0–415–14095–1 (pbk)

CONTENTS

PLATES AND FIGURES

PLATES

FIGURES

TABLES

ACKNOWLEDGEMENTS

•

I would like to thank Nick Entrikin, Linda McDowell, Carol Medlicott, Felicity Nussbaum and Gearoid O Tuathail for helpful comments and suggestions on an earlier version of this manuscript. Howard Harrington and Chase Langford gave timely assistance in producing the illustrations. Sarah Lloyd was a superbly supportive editor. Derek Gregory read the original manuscript with great care. His suggestions for rewriting were pivotal in making this book both more coherent and more accessible. I, however, remain solely responsible for what I have made of all comments and suggestions.

INTRODUCTION

•

In the 1990s such issues as the expansion of NATO membership to include former Communist states in Eastern Europe, the reintegration of Hong Kong into China after more than a century of British colonial rule, the proposed expansion of powers of the European Union at the expense of member governments and the threatened break-up of states such as Canada, Spain, Russia and the United Kingdom at the hands of regional separatist movements all combine questions of the organization of power with those of geographical definition. Understanding each of these issues involves paying detailed attention to the contexts in which they have arisen, knitting together accounts based on various local, regional and international causes. Since its invention in the late nineteenth century, however, political geography has addressed these kinds of questions largely by using an approach that reflected its origins in the Europe and North America of that time; by deploying the modern geopolitical imagination. This frames world politics in terms of a global context in which states vie for power outside their boundaries, gain control (formally and informally) over less modern regions and overtake other states in a worldwide pursuit of global primacy.

The purpose of this book is to identify the major elements of this approach and to show their historical–geographical specificity. In other words, the aim of the book is to show how political geography incorporated the dominant geopolitical imagination from the European–American experience that was then projected onto the rest of the world and into the future. Yet, today this approach seems to offer much less purchase on the emerging reality of world affairs than it once did. For example, as the Berlin Wall fell in November 1989, so also did the geopolitical template of East versus West that had underwritten world politics for the previous forty years.

As a result, these are anxious if exciting times, intellectually and politically, for political geography. A number of crucial 'certainties' from as recently as ten years ago, about, for example, fixed and unquestioned political boundaries between states, the division of the world into mutually hostile armed camps on the basis of political ideology, the centrality of states to world politics, and the primacy of fixed national identities in political psychology, have all either disappeared or are in question. The end of the Cold War, the growing importance of trading blocs such as the European Union, and the proliferation of ethnic and regionalist movements within established states have served to undermine the conventional wisdom. Perhaps it is not entirely surprising, therefore, that there is increased attention to the ways in which academics and political leaders have understood and practiced world politics. In times of flux, conventional wisdom is more open to scrutiny.

One angle of vision focuses on the role of geographical understanding of politics in the world today. The phrase 'world politics' itself conveys a sense of a geographical scale beyond that of any particular state or a locality in which states and other actors come together to engage in a number of activities (diplomacy, military action, aid, fiscal and monetary activities, legal regulation, charitable acts, etc.) that are intended to influence others and extend the power (political, economic, and moral) of the particular actors who engage in them. But the activities also rest on more specific geographical assumptions about *where* best to act and why this makes sense. The world is actively 'spatialized,' divided up, labeled, sorted out into a hierarchy of places of greater or lesser 'importance' by political geographers, other academics and political leaders. This process provides the geographical framing within which political élites and mass publics act in the world in pursuit of their own identities and interests.

The term geopolitics has long been used to refer to the study of the geographical representations and practices that underpin world politics. The word 'geopolitics' has in fact undergone something of a revival in recent years. The term is now used freely to refer to such phenomena as international boundary disputes, the structure of global finance, and geographical patterns of election results. One expropriation of the term ascribes to it a more specific meaning: examination of the geographical assumptions, designations, and understandings that enter into the making of world politics. Some recent work in political geography has tried to show the usefulness of adopting this definition of the term. In using this definition here, I attempt to survey how the 'discovery' and incorporation of the world as a single unit and the development of the territorial state as a political ideal came together to create the context for modern world politics. Studying geopolitics in this sense, or 'geo-politics,' so as to distinguish it from what is being described, means trying to understand how it came about that one state's prospects in relation to others' were *seen* in relation to global conditions that were *viewed* as setting limits and defining possibilities for a state's 'success' in the global arena. The use of visual language is deliberate. World politics was invented only when it became possible to *see* the world (in the imagination) as a whole and pursue goals in relation to that geographical scale.

According to the history of the word, geopolitics began in 1899 when the word was first coined by the Swedish political scientist Rudolf Kjellén. In so doing, thinking globally on behalf of particular states was explicitly connected by formal geopolitical reasoning to their potential for acting globally. Ways of thinking and acting geographically implicit in the phrase 'world politics' had begun much earlier, however, when the 'intellectuals of statecraft' of the European states (leaders, military strategists, political theorists) pursuing their 'interests' had to consider their strategies in terms of global conditions revealed to them by the European encounter with the rest of the world. The onset of the capitalist world economy and the growth of the European territorial state gave rise to a novel set of understandings about the partitioning of terrestrial space. The 'layering' of global space from the world scale downwards created a hierarchy of geographical scales through which political–economic reality was seen; in order of importance, the four were the global (the scale of the world as a whole), the international (the scale of the relations between states), the domestic/national (the scale of individual states) and the regional (the scale of the parts of the state). Problems and policies were defined in terms of the geographical scales (either domestic/national or international/foreign) at which they were seen as operating within a global context. World politics worked from the global scale down. It was at this scale, therefore, that the term geopolitics was usually applied. Yet it rested on assumptions about the relative importance of the various geographical scales to life on the planet that were already in place. The global and the national were privileged largely to the exclusion of the others. Naming such assumptions 'geopolitics' and creating models that claimed a superior grasp of geopolitical realities came much later, therefore, than the onset of the geopolitical representations and practices themselves. The purpose of this book is to identify and critically examine the central features of the geopolitical imagination that arose in Europe, beginning in the fifteenth century and that later became the stock-in-trade of political geography.

Two familiar examples from recent world politics can show how certain geopolitical understandings about geographical scale (representations) serve to underwrite specific 'policies' and 'interventions' (practices) that are then interpreted in terms of these understandings. The first example concerns the recent conflicts in the former Yugoslavia. The break-up of the former Yugoslavia along ethnic–regional lines in the 1989–1994 period might appear to have had *local* causes relating to the distribution of power and economic well-being of the various parties in dispute (Serb groups, Croats, Bosnian Muslims, others). But outside actors, such as the US and German governments, framed their policies largely in terms of bigger geopolitical pictures. These policies were crucial to the course of the conflict because all the internal parties needed external allies. For the USA, maintaining the territorial integrity of the country was paramount throughout the early years of the conflict because Yugoslavia was seen as a *buffer state* between the Soviet Union and its allies to the east and the USA and its allies to the west. The then West German government adopted a very different position. Thinking in terms of the investment and trade potential of the wealthier regions of Slovenia and Croatia, the German government encouraged political independence for both these regions and

Bosnia without requiring guarantees as to how minority populations (largely Serb) would be treated by the newly independent states. In both cases geographical framings that reflected different globally defined interests on the part of powerful external actors contributed to the conflict, even though each was rhetorically devoted to resolving it.

Many commentators have made much of the geopolitics of the break-up of the country, paying little or no attention to the nesting of causes across geographical scales. This reflects the privileging of the global and national scales implicit in the modern geopolitical imagination. At the same time, much of the discussion of maintaining the original state or recognizing the various parts was conducted in terms of the absolute sovereignty of individual governments to do this or that within 'their' territories. Statehood was the name of the game. The objectives of all sides were also expressed in terms of facilitating the entry of the area into 'the West,' so as to lead it out of its economic and political backwardness. Finally, the various external parties (the USA, the European Union, Germany, France, etc.) were all concerned about the fall-out from their policies on their respective positions in the global pecking order of states. For example, the US government worried that the former Soviet Union (as it was at the outset of the conflict) would gain political benefit from a divided Yugoslavia at American expense. Each and every one of these is a feature of the modern geopolitical imagination with a strong hold over political élites (and political geographers) around the world.

American intervention in Vietnam in the 1960s was also based on a similar set of geopolitical assumptions. The first was that US involvement was a necessary part of the geographical *containment* of world communism as represented by the Soviet Union and its allies, including the Viet Cong in South Vietnam and the Communist government in the north, as first enunciated in the years 1947 to 1950. The second, and inseparable from the axioms of containment, such as *domino* effects – as one country after another would fall if a stand was not taken in the first place – and *linkage* – Vietnam could be understood only in relation to the overarching global conflict between the USA and the Soviet Union – and *credibility* – if a stand was not taken there other (more important) allies would question American commitment to them – was that historical analogy pointed to the universality of these geographical 'truths.' American leaders used 'history' to justify their decisions, as if history simply repeated itself without the necessity of geographically informed judgment about the character of different world situations. The particular historical analogies were so familiar to the generation of American politicians who engineered American intervention in Vietnam that they could be communicated by single words: 'Munich' and 'Korea.' Each was shorthand for the future costs of appeasement of 'aggression' or intimidation. Only if a Viet Cong mobilization was contained could a future bloodier conflict be deterred. Finally, 'North' Vietnam was portrayed as 'invading' 'South' Vietnam, so as to define the conflict as the result of an act of international aggression rather than a civil war between competing political factions who at one time (during the latter phases of French colonial rule) had operated in the country as a whole. But the partition of 1954, achieved largely at the behest of the US government which refused to allow all-Vietnam elections after French withdrawal, became the crucial geographical representation in characterizing the war to the American population as a war of invasion that had to be resisted. In this account the Viet Cong was a 'fifth column' (a term first used in the Spanish Civil War of the late 1930s but going back to the Trojan Horse of Greek antiquity) acting as agents of the 'North' in the 'South.' Of course, as we now know, the 'loss' of South Vietnam following American withdrawal under heavy domestic political pressure did not lead to the Communist takeover of all Southeast Asia. Key geographical assumptions proved fallacious, yet they were very much the synthetic elements in lending meaning to the intervention as a whole, as painfully related in the memoir of the then-US Secretary of Defense, Robert S. McNamara (1995). More recent American military interventions, such as those in Granada, Kuwait, and Haiti, though differing in detail from that in Vietnam, have been accompanied by the same kind of geographical rhetoric, including frequent reference to the 'lessons' of Vietnam (restrict media access, bomb endlessly, avoid slow escalation, etc.) indicating the extent to which American foreign policy as whole became dependent upon a particular set of geopolitical assumptions during the Cold War period of 1947–1989.

These geographical assumptions and language indicate the facets of a more inclusive geopolitical imagination that insisted on dividing the world as whole into units of sovereign statehood, the challenge and response of a particular Superpower (the USA) to the pursuit of 'primacy' (becoming 'Number 1') by other leading states (such as the Soviet Union), the characterization of South Vietnam as a 'backward' country that could only 'develop' successfully by imitating the American example of 'modernity,' and the siting of Vietnam in a singular global context (that of the 'Cold War') rather than in terms of 'local' causes of political resentment and revolution. These fundamentals of the modern geopolitical imagination did not suddenly spring forth during the Cold War, whatever the specifics of the historical context. They had long-standing roots in the relationships established initially by European states among themselves and in their outlook towards the rest of the world. The first four chapters of this book are devoted to each of the 'fundamentals.'

An important caveat is that the geopolitical imagination as construed in the previous paragraph has never exercised *absolute* power over the course of world politics, in the sense of transcending the effects of technological, economic, and other material determinants, or in totally dominating understanding by political geographers and others. The modern geopolitical imagination has, however, provided meaning and rationalization to practice by political élites. It has defined the 'ideological space,' to use Immanuel Wallerstein's (1991) phrase, from which the geographic categories upon which the world is organized and works are derived. From this point of view, the history of modern world politics has been structured by practices based on a set of understandings about 'the way the world works' that together constitute the elements of the modern geopolitical imagination. It is Eurocentric because Europe and its offshoots (such as Russia and the United States) came to dominate the world. Political élites around the world adjusted to and adapted understandings and practices emanating from Europe. A large part of Europe's heritage to the world is the continuing strength of the 'common sense' about world politics bequeathed in the form of the geopolitical imagination. From Brasilia to Seoul and from Cairo to Beijing the dominant model is still that invented originally in Europe.

This does not mean that there have not been other 'causes' of world politics to which the geopolitical imagination has had to relate and adjust, such as technological and economic changes. For one thing, the rise of European power had its material roots based on, at minimum, economic and social pressures to expand geographically, military and naval technological advantages, diseases to which other people had limited or no immunity and an ability to borrow from and synthesize elements of other cultures. Neither does the importance of the rise of a modern geopolitical imagination mean that 'oppositional' ideological currents (such as anti-colonialism, *laissez-faire* economics and some varieties of socialism, for example) have not challenged the dominant representations and practices; only that these have had to contend with a powerful and well-established conventional wisdom about 'how things are' that has reproduced the state-based and place-hierarchy view of the world that lies at its heart. Movements to remove colonial rule often ended up accepting and sometimes celebrating the often problematic political boundaries imposed by colonial empires. India and the countries of Africa are classic cases. 'Socialism in one country' (as in the former Soviet Union or in contemporary China) and neo-liberal economic polices based on activist governments (as with the Reagan administration in the USA, 1981–1988, and the Thatcher government, 1979–1990, in Britain) reflect the channeling of oppositional discourses within parameters defined in large part by the hegemonic geopolitical imagination.

The dominant representations and practices (or hegemony) constituting the modern geopolitical imagination have been overwhelmingly those of the political élites of the Great Powers, those states and empires most capable of imposing themselves and their views on the rest of the world. Membership of this group depends on recognition by existing members. Qualification has not depended simply on coercive power, the ability to force others to do what you want, but also on the capacity to write the political–economic agenda of others, defining appropriate standards of conduct and providing the framing for inter-state relations with which others must conform if they are to gain recognition and rewards from the Great Powers. News stories using the 'common sense' accounts of what is at stake for 'us' or others in this or that part of the world, 'official stories' told by political leaders, and the presentations of intellectuals elaborating on the 'logic' of particular foreign policies and military strategies are all important ways in which the dominant story-lines and agendas of different states (and other actors) can be disseminated, both domestically and internationally.

At any one time a single state can be *number 1* among the Great Powers; having a key role in laying down and enforcing the 'rules of the game' (often referred to by the term 'hegemony'). This was the case with Britain in the mid-nineteenth century and with the United States from 1945 to the 1970s. But as long as political élites engage in relations with one another there must be rules governing their interaction. So, even in the absence of a particular hegemon there is still hegemony in the sense of a set of dominant understandings and practices regulating world politics. Recently, a transnational liberalism sponsored by governing élites around the world seems to offer an emerging ideological hegemony without a single hegemon (or dominant state). Based more on principles of free trade and comparative advantage working to the benefit of global business enterprises (and their local allies) than on the enhancement of state power, its existence calls into question many of the established tenets of the modern geopolitical imagination. Its emergence does not signal the 'end of geopolitics,' however, so much as its possible reformulation from a state- towards a firm- and city-centric focus in which states must adjust to new demands on them to relax their controls over markets. As yet, however, the modern geopolitical imagination, though changing its emphases as the balance of power among the Great Powers and the nature of the world economy have changed over time, still remains prevalent in framing the conduct of world politics. Not surprisingly, political geography and adjacent fields of study such as international relations and political sociology remain largely attached to the same vision. What are the roots of this imagination? Why has it had such staying power?

The first four chapters focus on key 'principles' of the modern geopolitical imagination and how it has developed down the years. The first chapter addresses a primary feature: a global vision without which 'world' politics would be impossible. The acquisition and perpetuation of this vision – the sense of a world-as-a-whole that powerful actors must survey and subdue – are traced through the history of early modern cartography, the theorizing of 'one world of humanity,' imperial cosmologies, colonial economics, formal geopolitical models and the global ideological polarization of the Cold War. The purpose is to show the cumulative historical basis to global geographical visualization: how the Earth was made into the World. Thinking about the world as a whole was not a one-time thing, established once and then taken for granted. It had to be constantly reproduced in changing economic and technological circumstances to remain effective. Seeing the world as one and then dividing it into a hierarchy of places have required relating thinking about the world to changing material conditions even as the *a priori* assumption of wholeness is perpetuated.

Encapsulating the world in the mind's eye as the framework for understanding anything that happens anywhere also can serve to legitimize what could be a very partial view of the world, from one's own position at a particular location, into a view that could claim to be a 'view from nowhere:' an objective view of the world as from outer space. This is not to say that such a view is *a priori* illegitimate intellectually and politically, only that the association of global geopolitical thinking with the self-defined 'interests' of specific states (such as Germany or Britain) tended to use the scientific claim to objectivity on behalf of a particular identity/interest. This use of the view from nowhere has always been an important part of the modern geopolitical imagination. Its relating of anywhere to everywhere rests on the legitimacy derived from its being a view from nowhere. If it was seen as just a view from somewhere, its partiality and situatedness would be signaled from the outset. In this way a political claim on behalf of a particular interest is turned into a natural claim about the world as it is.

A fundamental feature of the contrasts drawn between different parts of the world seen as a whole has been the labeling of 'blocks' of global space as exhibiting the essential attributes of the previous historical experience of the dominant block. This translation of 'time into space' is the subject of the second chapter. Typically, modern geographical taxonomy involves the naming of different world-regions or areas as 'advanced' or 'primitive', 'modern' or 'backward.' Europe and some of its political–cultural offspring (such as the United States) are seen as defining modernity and other parts of the world only figure in terms of how they appear relative to Europe's past. Being like or imitating Europe thus becomes a condition for entry into the state system (as opposed to justifying subjugation) and provides the norm or standard for judgment about particular states (who is most advanced?, etc.) The case of the idea of Three Worlds of Development will be used to show how the global partitioning of space according to temporal 'stages' of development has become woven into world politics. The exploration and early settlement of Australia are used to illustrate the ways in which an existing understanding of that continent's place in the world informed the outlook and behavior of the explorers and settlers. The case of the historiography of Italy, a European state often seen as 'lagging behind' the rest of the continent, illustrates how pervasive the use of 'time into space' judgment has been in the making of the modern geopolitical imagination.

For world politics the fundamental world map is a world political one: the map of territorial states. The third chapter identifies the geographical assumptions that produce this state-centric aspect of the geopolitical imagination. Three assumptions are seen as integral to the state-centricity of the modern geopolitical imagination: first, state sovereignty and territorial space; second, the territorial state as container of society; and finally, the domestic/foreign polarity. These assumptions are modern ones; the first dates from the sixteenth century at the earliest and the others are even more recent. Yet, the idea of a world made up entirely of states or territorial actors, as opposed to other forms of polity, is given a transcendental force within the geopolitical imagination. In world politics space can only be divided between actual states and states-in-the-making. Other modes of organizing politics geographically (such as kin-based polities, city-states, classic empires such as those of Ottoman Turkey or Manchu China, and confederations) are seen as either relicts or 'really' territorial states in disguise (they can be regarded 'as if' they were real states). Territorial states are the individual actors of the geopolitical imagination. This perspective, cleaning up the large variety of polities to be found in all epochs into a single type, is often referred to as the 'Westphalian view,' after the Treaty of Westphalia (1648) which settled the religious wars in Europe by establishing as a standard the idea of a single governmental jurisdiction over a single territory.

The dynamic force bringing together the other elements of the modern geopolitical imagination is devotion to the idea that pursuit of one state's (typically, yours) interests or 'security' relative to those of all others is necessary for personal ontological security. The modern world is seen as one of unremitting competition for primacy: either to dominate the world economically (in some recent understandings) or to turn it into a world empire (in more typical usage). The fourth chapter addresses three questions that arise from this conception of world politics: What makes a Great Power a potential 'hegemonic' state? This directs attention to two crucial features of thinking about inter-state competition: uneven rates of economic growth and international anarchy. Two other questions follow: Under what historical–geographical conditions has the pursuit of primacy been possible? And, is it still possible at the outset of the twenty-first century?

Finally, how have the geopolitical underpinnings of modern world politics changed over time? Chapter 5 provides a narrative account of three 'Ages of Geopolitics' in which the modern geopolitical imagination has shown distinctive features and relations to practice. The first, prevalent in the eighteenth and early nineteenth centuries, was a *civilizational* geopolitics, in which Europe's unique civilization compared to the newly discovered 'rest of the world' played a central role. The second, dominant from the late nineteenth century to 1945, the epoch when political geography as a field was established, was a *naturalized* geopolitics, in which the 'natural' character of states as predators and competitors assumed a key position. The third, operational during the years of the Cold War, an *ideological* geopolitics, was based on dividing up the world between competing ideas about how best to organize political and economic life ('socialism' versus 'capitalism,' etc.).

Although the 'principles' of geopolitics as laid out in Chapters 1 to 4 remained in effect, the net impact was not always the same. Within a general continuity, therefore, one can identify distinctive epochs in which the geographical representations and practices implicit in world politics have undergone important shifts. A final section in Chapter 5 considers what recent great political changes, such as the end of the Cold War, and technological and economic changes, such as those associated with an emerging informational capitalism, might augur for the future of the geopolitical imagination and for the political geography that remains attached to it.

The concluding chapter provides a theoretical review of how the first four chapters relate to the fifth and uses the arguments of previous chapters to claim that political geography stands in need of re-composition beyond the frame of reference supplied by the modern geopolitical imagination. This new approach would resist privileging certain geographical scales of analysis, such as the global and the national, as supposedly of singular importance. Rather, rethinking geographical analysis would lead to an emphasis on how different scales relate to one another differentially over time. This leads away from state-centered political horizons towards a more pluralistic vision of political organization in the future.

By way of summary for the book as a whole, world politics can be analyzed in terms of a set of principles that collectively constitute the modern geopolitical imagination. But these must also be put together in terms of how they have operated in different historical epochs. By using an analysis of the modern geopolitical imagination as a way of interrogating world politics, therefore, a new narrative structure can be given to modern world history (Chapter 5), in the process also identifying the critical geopolitical understandings and practices upon which world politics and many approaches to understanding it, like political geography, continue to rely (Chapters 1 to 4). That it is no longer sufficient and needs to be replaced by a different approach to understanding the world's political geography is the implicit theme of the book as a whole.

1

VISUALIZING GLOBAL SPACE

•

An account of the modern geopolitical imagination, the predominant ways world politics have been represented and acted on geographically by both major actors and commentators over the past two hundred years, must start with the origins and development of the capacity to see the world as a whole. From this point of view, the 'modern' world is defined by the imaginative ability to transcend the spatial limits imposed by everyday life and contemplate the world conceived and grasped as a picture. The geopolitical imagination, therefore, is a defining element of modernity. Its most distinguishing feature is the conception of the world as a single if divided physical–political entity; a feat of imagination impossible before the encounters of Europeans with the rest of the world beginning in the late fifteenth and early sixteenth centuries.

Two characteristics of visualizing global space emerged at the outset of the European Age of Discovery. Ever since, they have been reproduced in the governing principles of geographic thought and through the practices of statecraft. The first, and the one that has received most recent comment, is that seeing the world-as-a-picture, as an ordered, structured whole, separates the self who is viewing from the world itself. The observer stands outside of terrestrial space, so to speak, and frames the world as apart from and prior to the places and people it contains. This seems to be a peculiarly European perspective in origin; associated with the Renaissance-era separation of the observer from the world and an approach to knowing that insists on privileging vision as the most 'noble' of the human senses. What is 'seen,' even from over the horizon by means of such tools as world maps, is what exists. The map is an accurate report of what is there. Representation and world are as one. In reaction to this ennobling of singular vision and the absence of skepticism about its impact, much recent philosophy has become deeply suspicious of visuality and what Martin Jay (1993, 14) terms 'its hegemonic role in the modern era.' This doubt can be traced to the philosopher Friedrich Nietzsche's (1844–1900) and more recent feminist and other critiques of the 'view from nowhere;' there can be no pure, will-less, timeless, placeless knower, no pure reason, no absolute knowledge or absolute intelligence. All seeing and knowing is a perspective, drawn from a situated point of view. From this contrarian standpoint, real objectivity lies not in the promulgation of a single perspective from within a singular historical experience but from the introduction of as many eyes (perspectives) as possible. There is never a single view from nowhere.

The second characteristic of global visualization is that the world pictured beyond the horizon is a source of chaos and danger. The evil spirits and dark places decorating the borders of early modern maps, where they signified 'unknown' and presumably dangerous places, moved into the world itself where they came to represent fearful religious, civilizational, and political differences. The one world-picture, therefore, is not a composition of equal and pacific elements but a hierarchy of places, from known to unknown, from most friendly to most dangerous. The most well-known representation of this character is that of a dichotomous global West and East, in which the former is seen as the total opposite and, hence, definitive standard for the latter. With roots in an ancient European past, this opposition serves as a geographical template onto which more local differences can be mapped. They then become explicable only as elements in the bigger picture. Local differences are ascribed to worldwide distinctions rather than to local differences *per se*. The local has meaning only in relation to the global. Without the global reference, difference cannot be articulated. This is the attribute of linkage whereby specific places are drawn into an overarching global geographical frame of reference.

This chapter is about the origins and elaboration of the twin features of global visualization. I begin with the Renaissance origins of the view of the world from beyond its limits: the view from nowhere. A case is then made for how this perspective (in both theoretical and visual senses of the word!) was both elaborated on and familiarized by the empiricist conception of knowledge and its commitment to vision as the primary human sense. With respect to the second feature, I identify some of the sources of binary geographical thinking, particularly the East–West opposition, and argue for how these have been reinforced historically by means, for example, of imperial cosmologies, colonial economics, formal geopolitical schemas, historical analogies, and informational technologies. Coming to see the world as both a whole and as dangerous was not a one-time thing; neatly packaged at one moment and thereafter reproduced in exactly the same way without ambiguity or challenge. Visualizing global space has involved the adaptation of its two features within the overall context of a persisting modern geopolitical imagination.

SEEING THE WORLD AS A STRUCTURED WHOLE

Events more than ideas capture immediate popular attention. So it was neither Galileo's refutation of geocentrism, that a stationary, immovable earth was the center of the cosmos, nor the anonymous discovery that the Sun is a star that produced the realization that the earth is a rotating spherical object on whose surface everywhere is connected to everywhere else. It was perhaps the return of Magellan's crew in 1522 from the first known circumnavigation of the earth that really brought this home. The Portuguese explorer never made it back himself. He was killed in what later were named the Philippines (after King Philip II of Spain). 'The Europeans,' to quote J. H. Elliot (1991, 10), 'had found space, and found it on an unimagined scale. But, paradoxically, even as their world expanded, it also began to shrink. A globe encompassed became a globe reduced.' Thereafter, Europe was no longer the world and the world was no longer the center of the universe. This had revolutionary consequences for both the European world view and the meaning of life:

> Since the earth was revolving daily, heaven and hell could not be located where they had been thought to be, and in rational minds there was a growing skepticism that either of them existed. Satan without hell was implausible. God without heaven was inconceivable, at least the medieval God was, but here reason ended.
>
> (Manchester 1992, 295)

The familiar vertical conception of the universe (the Great Chain of Being) that connected ordinary mortal Europeans into the universal scheme of things was challenged by an opening up of horizontal horizons that offered a new vision of the world and Europe's place in it. An alternative framework for understanding the new world was needed and, in basic outline, it was fast in coming.

The Renaissance rediscovery of Ptolemy (who lived AD 90–168) provided a suitable model of the structure of the world. Rehabilitated initially by Sebastian Munster, Ptolemy's cosmography (an imagined world structure into which phenomena for study in their own right can be slotted) supposed a global world with no limits other than those of its poles, regions, and zones (Figure 1.1). This model was inadequate in many of its details, as would be revealed over the next several hundred years. For example, in its account the glacial and torrid zones were uninhabitable by humans, there was exact symmetry between the northern and southern hemispheres, and America had no place; stretching as it did from north to south and in two blocks joined by a land bridge, it was also organized in a very different way from that of the Old World's division into the three continents of Europe, Asia, and Africa. But Ptolemy's model offered a ready substitute for the medieval world view. Its main advantage was its openness to the filling in of the 'unknown' spaces as more voyages led to the enlargement of perspective and the impetus for trade and conquest prompted speculation and inquiry about 'unvisited' parts. Most of all,

> this model, apparently so inadequate, proved to be fecund by virtue of its own anachronism. It offered to modern geographers a three-quarters empty canvas, leaving them free to

inscribe on it the delineation of newly 'invented' or discovered lands; a form, at once closed and open, full and lacunary, that represented the ideal construction in which to house, with their approximate and disparate localizations, the 'bits' of space that navigators brought back from their distant voyages.

(Lestringant 1994, 7)

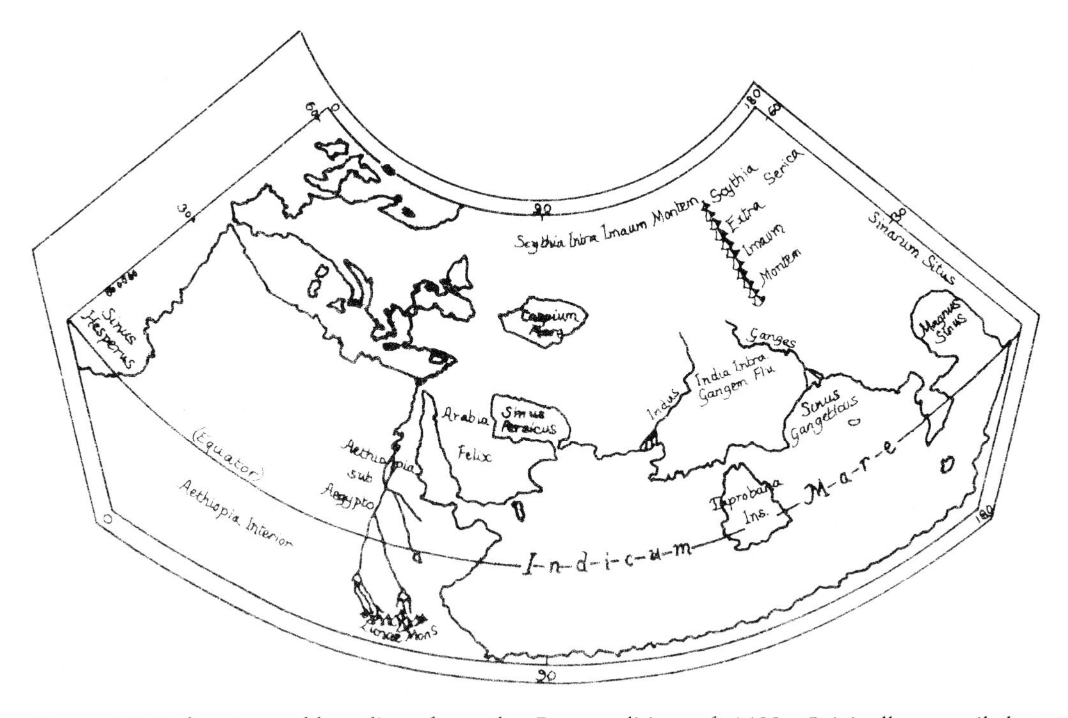

Figure 1.1 Ptolemy's world outline, from the Rome edition of 1409. Originally compiled at Alexandria, Ptolemy's guide to geography was preserved for posterity by Arab and Byzantine scholars. Translated into Latin from a Greek manuscript in Florence in 1409, printed editions only circulated widely in Europe in the last decades of the fifteenth century
Source: Crone, 1978.

On this open 'canvas' could be played out the 'cosmographical fictions,' such as the search for the north-west passage or a southern continent, that inspired so many of the personal and national projects of exploration and conquest.

The expansion of speculation and knowledge about the world on a Ptolemaic basis is most clearly recorded in the world maps of the sixteenth to nineteenth centuries. In the early sixteenth century Ptolemy's original maps were generally predominant in Europe. Very quickly, however, national rivalries underwrote the efforts of cartographers to keep up with (and sometimes to anticipate!) the process of discovery. As early as 1489 Henricus Martellus, though still following Ptolemy, introduced a new coastline for South Africa to take into account the voyage of Diaz three years previously. By 1506 Giovanni Contarini showed a complete outline for Africa and noted the discoveries of Columbus. In 1529 Diego Ribeiro made use of Antonio Pigafetta's report of Magellan's journey to construct a new world map. Finally, Blaeu's map, published in the Netherlands in 1648 to celebrate the Peace of Westphalia and symbolically representative of the state-centered world then in ascendancy, showed much of the outline of Australia and New Zealand as reported by Abel Tasman in his voyage of 1642–43. Thus, by the middle of the seventeenth century European scholars and political leaders already knew that they lived in a world that had a definite shape, even if actual knowledge of all of it was still incomplete by twentieth-century standards.

These maps, speculative as they often were, share two features. One, common also to small-scale maps in previous and later epochs, was a pervasive ethnocentricity. The maps are centered on Europe (notably in the long-popular Mercator map), conveying a clear sense of Europe's ideological centrality. There is nothing passive about a map or a compilation of maps in an atlas. They can convey a strong message on behalf of a particular 'world view.' For example, eighteenth-century British maps of the American Colonies included insets and cartouches that vilified the French, stereotyped African slaves, removed Indians cartographically before they could be removed physically, and showed the extent of British dominion over an inviting landscape. The second feature is their hierarchical representation of space, identifying and naming sites in terms of social and geopolitical significance. The power of the maps lies in their masking of these features behind a veneer of objectivity. The selectivity that went into the construction of the maps is never made evident. The reader has to presume a fairly close approximation between the maps and what lay 'out there' in the world beyond immediate experience. The maps convey an image of a world without the actively mediating hand of the cartographer. What appeared was surely what was there?

The modern cosmographer had a number of differences from the ancient one, therefore. The first was with respect to the perception of space. Perception of space was no longer totally abstract. The horizon could now be crossed. This gave a sense of both ubiquity and omnipotence that Ptolemy had lacked. The world now could be both thought of *and* experienced as a whole. Indeed, the practice of navigation over the open sea required a precise positioning of the traveling self in relation to the world as a whole. The second was the increasingly direct experience of the world which gave travelers the authority to speak about where they had been in terms of their determination of its place in the world. Having been somewhere now provided a license to speculate about everywhere. The ancients could only conjecture; the new knowledge of the world was based on an amalgam of personal vision, fantasy and speculation. So while drawing on ancient Greek sources for its model, modern cosmography combined them with both the astrolobe and the globe of the navigator and the naïve experience of the observer–writer. The third, reflecting the etymology of the Greek word *kosmos*, the root of both cosmography and cosmetic, was the aesthetic attention given to the singularity or beauty of objects. At the start, as in the case of the French cosmographer Thevet, it was only the uniqueness of what was discovered in terms of its absolute empirical variety that mattered. The wonder and novelty of the world out there and the sheer amount of new information sometimes resisted open

Plate 1.1 A German woodcut of *c.*1505 showing a representation of the Topinambas of coastal Brazil. The picture combines a confusing mixture of routine cannibalism and domestic life which suggests that its creator was not quite sure what to make of stories about the Topinambas and erred on the side of including everything he had heard

Source: The Spencer Collection, New York Public Library

theoretical classification (Plate 1.1). But with the rise of specific disciplines devoted to this or that phenomenon, order was imposed on the immense variety of the world by drawing on what was thought of as the experience and example of Europe itself. The form that this took is examined in Chapter 2.

The cosmographical model, the projection of a presumed structure to the world's physical geography drawn from ancient sources, was not to last in unrefined form. Very quickly more specialized studies of Geography (large-scale global patterns of seas, continents and climates) and Chorography (local regions and landscapes) replaced the focus on the earth within a larger universe implied by Cosmography. But it had provided an orientation to the world as a whole upon which later generations were to build. Travelers carried with them its view of 'spaceship earth' even as their adventures slowly undermined many of its geographical presuppositions. In the first place, it replaced the Creator with the 'seeing' observer who in the name of experience spoke from authority. It established the view from nowhere as the modern ideal. It also used partial knowledge (fieldwork?) to justify knowledge of the whole. Having been somewhere was equivalent to having been everywhere. Images rather than words, or what Cosgrove (1996) calls a 'graphic imperative,' gave authority to both the Geography and Chorography that descended from the older Cosmography. Finally, the cosmographical model saw the world as a whole; a canvas upon which a modern, Europe-centered world could be painted. Finding out was not so much about understanding in local terms as about compiling and collecting so as to incorporate the world's variety into categories familiar to the European mind.

The Renaissance, however, had another influence beyond that of the revival and elaboration of Ptolemy's cosmography. This was the discovery of 'perspective'. Brunelleschi, the great architect of the Duomo in Florence, is technically the author of that optic of artistic representation in which a structure or a scene is viewed in a linear plane which the architect or artist uses to communicate to the observer an intended meaning. Thus a consistent reading of a building or a painting is made possible by imposing an orderly vision by means of an external point of viewing from which an object always appears in the same light. This conception of the consistent gaze as the key to understanding was not restricted to the pictorial arts. The linear perspective was an integral part of the Renaissance 'world view' (!). The language of knowing – perspective, view, vision, world view – was expressed in overwhelmingly visual terms. The world was an object separated from the viewing subject, that could only be understood adequately if seen as a whole. Consequently, 'parts' are understood only in relation to the whole. Perspective allows a framing or 'field of projection' of particulars as elements in an ordered whole.

John Berger (1972, 16) has summarized the implications of the rise of perspective as the governing principle for both viewing and knowing:

> The convention of the perspective, which is unique to European art and which was first established in the early Renaissance, centers everything on the eye of the beholder. It is like a beam from a lighthouse – only instead of light traveling outward, appearances travel in. The convention called those appearances *reality*. Perspective makes the single eye the center of the visible world. Everything converges on the eye as the vanishing point of infinity. The visible world is arranged for the spectator as the universe was once thought to be arranged by God.

The idea of a uniform space falling away from the observer underwrote this vision. Within it a visual field would have definite limits, but what is seen would always be conceptualized with reference to it. This fit into the evolving economy of the times. Objects, such as those in a visual field, had no intrinsic value outside of their relations to that field; equivalent to the exchange value of goods in capitalist trade. The rise of perspective, in liberating knowing from obedience to past texts, therefore, also brought with it an orientation towards objectification of the world in the interest of trade and commerce. This allowed the world to be divided up into a variety of grids and reproduced using a number of projections. This quantification of the world into precise bits and pieces (by latitude and longitude, for example) was possible only because of the visualization of the world as whole. Powers of 'planetary scope' could henceforth be deployed to control sea lanes and routes of circulation. In this way global space was 'hierarchized' and ordered into zones of greater and lesser significance in which different activities and behaviors were appropriate. Technical systems for ordering the earth's surface, therefore, were important elements in the development of the modern geopolitical imagination.

Because it is now so familiar, the perspectival conception of knowing seems unremarkable. Yet, it involved a revolutionary shift in consciousness about the relationship of self to the world (and others), even if it had earlier roots. The self was now both external (as observer) but also in the world (as actor). To operate as the latter required the sense of the former. Previously, this separation between knowing and doing did not exist. Rather than based on the particulars of local knowledge, modern knowing required a prior framework into which knowledge about people and places could be pigeon-holed. Order could be discovered through creating distance between observer and observed and then seeing what is observed in relation to a whole setting. The European observer, trading and conquering as well as looking, saw the world 'as a differentiated, integrated, hierarchically ordered *whole*' (Gregory 1994, 36), in which he (it was usually a he) was both the external arbiter of truth about the nature of the parts and the primary agent in creating an integrated world where none had existed before.

The standard scientific model of knowing was built on this perspectival basis. Skepticism about the understanding to be gained from an explication of texts or great books (perhaps emanating, ironically given the scriptural claims on which it was based, from the Protestant Reformation's questioning of traditional authority) led to an emphasis on the virtue of 'direct observation' of facts that could be mapped and then subject to independent verification. An ethic of cumulative achievement of accuracy removed the inevitable selectivity of all maps or other representations of the world (however sophisticated) from active consideration. Rather than simply mirroring the world, therefore, popular maps helped to constitute it. In masking their selectivity behind empiricist claims to accurate representation they provided a powerful means of picturing the world as a whole as if it existed independently or separately of all attempts to conquer, tame, and exploit it.

There was no inevitability about drawing the connection between wholeness and hierarchy, however. Seeing the world whole, however selective in practice, could imply a sense of common fate, of a common humanity drawn together in planetary harmony, whatever local differences there might be. This was indeed the 'vision' of some Europeans. The Enlightenment cosmopolitanism of the German philosopher Herder (1744–1803) represents, for example, a version in which there is only one human race and a single human reason; climate and other geographical features are taken to account for such differences as exist between cultures. In this understanding, there can be no innately superior cultures and no groups or nations whose members are more fully 'human' than others. But European experience still provided the standard for what 'one humanity' might entail. Indeed, European experience defined what 'humanity' could be: a singular human nature resting on 'universal' attributes that were more often than not projections of European ones.

In the end more hierarchical positions prevailed, ones more in tune with the growth of capitalism and the development of the territorial state. Increasingly, uneven development following the advent of a world market and European industrialization, the pairing of modernity with backwardness in understanding how this occurred (see Chapter 2), and the definite hierarchy among states (see Chapters 3 and 4), acted to reinforce the sense that places were hierarchically ordered on a world scale. In particular, the nineteenth-century Romantic reaction to Enlightenment rationalism added to the older perspectivalism notions of 'national character' which claimed superiority in human status and viewing capability for particular European nationalities. As Britain became the world-hegemonic power, English 'common sense,' based on empiricist reverence for the accuracy of representations (such as maps) in directly revealing reality beyond the horizon, was established as the standard for judging 'things as they are.' Technological innovations such as the steamship, the telegraph, and the telephone and associated techniques for ordering the world into time-zones (to make possible train and steamship schedules) reinforced the sense of a 'closing world' which could be represented and understood in a straightforwardly empirical manner.

Knowing by seeing the world as a *horizontal* but hierarchically organized whole was finally institutionalized as the modern alternative to the *vertical* Great Chain of Being (connecting the supernatural to the human worlds) that had been the dominant cosmology of other older civilizations. The particular content might differ as intellectuals of statecraft in different states fused their own national myths into the general fabric of the geopolitical imagination. But the tendency to consider strategies and options within a whole-world framework was well established by the nineteenth century across all of the Great Powers and also those states which aspired to such status.

BINARY GEOGRAPHIES

It is well known that all societies define boundaries between themselves and others. The 'world' beyond the horizon is threatening as well as alluring. The zone of difference and danger is both repelling and attracting. At the same time, societies can only exist by defining themselves against an external standard; an Other without which the Self could not see itself as distinctive. 'European' society, the idea of Europe itself as a coherent social–geographical entity, could only arise in reference to what it was not and in relation to where it started and ended. Within Europe, and likewise, the nature of different national societies could only be known with respect to others they were not. Not all societies, however, are capable of imposing their maps of difference and danger on others. This is where material resources come into play. Europeans have been able on a world scale to do what only great empires, such as the Roman and Chinese, did more parochially: impose geographical boundaries on everyone else for a long period of time. In this way local differences have been invariably assimilated into a global geographical taxonomy with its roots in Europe.

The most basic division was between Europe and the other continents. The oldest surviving map of the world (from the seventh century AD) identifies three continents, Asia, Europe and Africa, does not have a surrounding ocean (as older ones probably did) and shows a large uninhabited area to the east. The boundary between Africa and Europe is the Mediterranean, that between Africa and Asia the River Nile, and that between Europe and Asia the River Tanais (the River Don) and the Meotides Paludes (the Sea of Azov). The map is, in the literal sense, 'oriented,' aligned to the east, where the sun rises and where Paradise was supposedly located. The idea that the sons of Noah populated the three continents after the Great Flood is reflected in the naming of each on the map. Shem receives Asia, Ham Africa, and Japheth Europe. Thus, the Jewish and Christian story from Genesis of what happened after the Flood is incorporated into the map with Japheth (the precursor of Christian Europe, according to the dominant commentary of the time) identified with the Christian continent. This theme was muted, however, until the fifteenth century when the identification of Europe with Christendom became usual. At this time the boundary between Europe and Asia moved considerably to the west. This was the result of divisions within Christianity (the Orthodox–Roman split), the resurgence of papal authority in western Europe following the Great Schism of 1378–1418, and the threat posed by the expansion of the Ottoman Turkish Empire. These conspired to produce a 'civilizational' definition of Europe in which the term *Respublica christiana* was applied to Europe and the adjective 'European' was used for the first time (both by Pope Pius II, 1458–64).

This sense of cultural unity was reinforced by two related intellectual trends. One was the spread of an 'ideal' of education based on reading Greek and Latin (so-called classical) authors in the original languages. This humanistic education linked the élites in European countries at the same time they were becoming politically divided by the emerging merchant capitalism and state formation of the time. The second was the tendency to reach back to the Greeks and Romans to give Europe a genealogy upon which its present unity could be seen to rest. Though the Europeans of the fourteenth and fifteenth centuries were largely descended from the tribes who had swept into Europe at the time of the collapse of the Roman Empire and until the Renaissance they had lost touch with the ancient world, they eagerly expropriated the 'Classical' past for themselves. A popular way to understand the new worlds that Europeans encountered from Columbus on was to assimilate them to Europe's 'own' unitary past; to the movement from pagan and barbarian to Christian and civilized. It was Rome, above all, that provided the political model and language for the developing European empires of Spain, Britain, and France. This 'imaginative dependence of the new upon the old' (Pagden 1994, 12) helped define what an empire might be and also provided a common set of understandings around which European imperial competition was organized. Each of the European empires could see itself as the inheritor of the imperial mantle of Rome, at the same time as their common competition restricted the main players to Europe. The imitation of Rome (if not of Christ) was possible only within the boundaries of Europe. The classic division between Europe and Asia reaching back to the Greeks, therefore, was recycled to give the concept of Europe a degree of independence from that of Christendom.

Renaissance political theories reinforced this division between Europe, Africa and Asia. One was in the invention of the concept of 'balance of power' which was held to apply, from the time of the Florentine political theorist Machiavelli (1469–1527) onwards, only to the balance between states in Europe. Another was the total opposition drawn between European political practices and those elsewhere. In particular, Asia or the 'East' was seen as despotic and lacking in pluralist forms of political organization. This, despite the historical record, as recounted by Patricia Springborg (1992), of a much more entrepreneurial and ungovernable East compared to a quiescent West lacking in much popular participation in politics until the twentieth century (also see Chapter 2). This historical inversion, according to Springborg (ibid., 20), can be understood as 'staking out territory and hoping, thereby, to create facts.' This point is also made somewhat more generally by Edward Said (1978, 54) in his book, *Orientalism*, when he argues that Europeans defined themselves (as do all groups, one might add) negatively, against others whose nature was largely unknown:

> the sense in which someone feels himself to be not-foreign is based on a very unrigorous idea of what is 'out there,' beyond one's own territory. All kinds of suppositions, associations, and fictions appear to crowd the space outside one's own.

The World of the East that Said identifies as crucial in defining the most important Other for the creation of Western self-images was in fact little known. This just goes to prove that a geographical imagination need not have much empirical content (Figure 1.2).

Christianity played a continuing role in the self-image of Europeans but by the eighteenth century it was no longer the dominant force it once had been. By then the growing perception of Europe as a center of artistic and intellectual invention had combined with the palpable sense of material progress to produce an awareness of

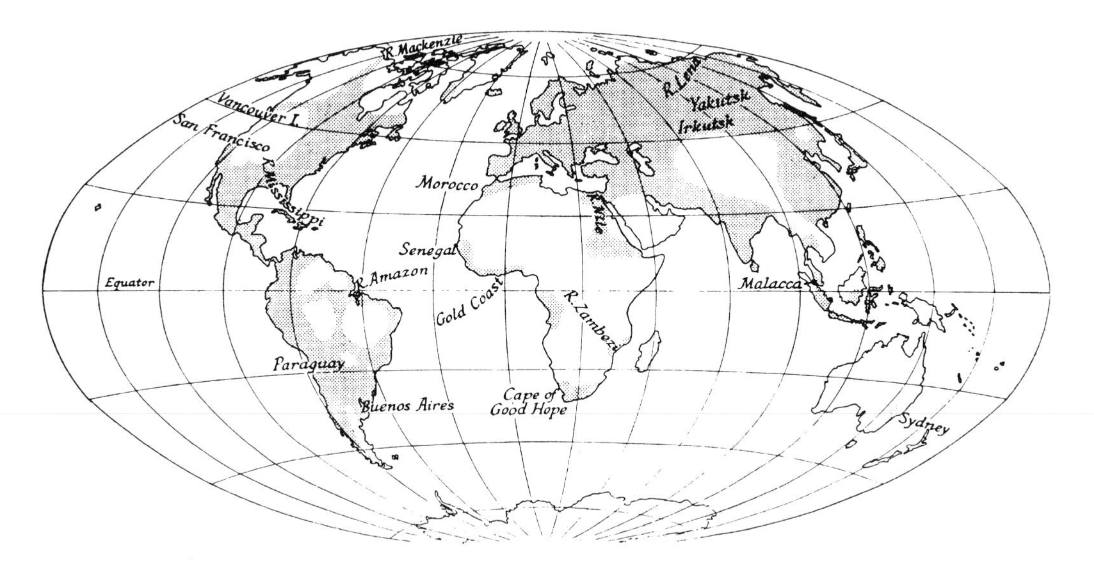

Figure 1.2 The world as known to Europeans *c*. 1800. Shaded areas represent those parts of the world known 'in some detail' by that date
Source: Crone, 1978

Europe as a new civilization. This did not lead simply to a growing feeling of European superiority. It also engendered the idea of 'levels' of civilization with Europe at the top but with the possibility that other regions might achieve greatness by following in Europe's footsteps. Other places and people could be understood satisfactorily only in relation to this global cultural hierarchy. The one big division, between East and West, led to smaller ones, particularly as local differences were discovered in the course of exploration, trade, and conquest. A whole fund of taxonomic lore developed, separating races, regions, and nations from one another in terms of their relations to the most abstract global distinction. Even within Europe, the distinction between Eastern and Western Europe was important in delimiting where Europe 'ended' and Asia 'began.' Not only the most 'exotic' regions, such as the South Pacific or Brazil, provided grist to the intellectual mill of differentiating Europe from the Rest. Racial and cultural classifications drawn from 'Eastern' Europe were, if anything, more important given the centrality of the Europe/Asia divide to all other geographical divisions.

The idea of race was vital to the substantive distinctions drawn between different parts of the world. There is little or no evidence of a conscious idea of 'race' before the sixteenth century. Its 'high point' was not reached until the late nineteenth and early twentieth centuries when the fine distinctions of 'scientific racism' replaced older, more informal means of drawing racial boundaries. In place of the uncertain stories about Shem, Ham, and Japheth, humankind was ordered after the 1680s according to differences of anatomy, climate, and pigmentation. Ironically, at first the old biblical division reappeared in which Northerlings, Middlers, and Southerlings substituted for the three sons of Noah, if now on natural rather than theological grounds. Later taxonomies departed from the tripartite formula, tying physical appearance to a large number of cultural characteristics. The different races, however, continued to be clearly associated with different world regions and with different stages of economic and moral development.

Within the broad racial divisions of humanity, finer racial differences were identified that served to explain 'national' origins in racial terms. For example, the French philosopher Montesquieu (1689–1755) remade the reputation of the northern barbarians who heretofore had been thought of as the brutal destroyers of the Roman Empire, by conceiving of them as one of the 'races' of time and order, rooted over centuries in climate, land, and soil. This served, and has served down the years since, to legitimize 'France' in a racial past independent of that of Rome. Indeed, disputes over this claim, and the counter-claim that France's origins lie in an older but derivative Gallo-Roman world, still animate French politics. In the autumn of 1996, for example, the Pope's arrival in France to celebrate the fifteenth-hundredth anniversary of the Christian baptism of Clovis, the King of the Franks, resurrected old arguments about French racial origins, including claims that Clovis either was or was not the 'first Frenchman.'

The linkage of the local into global and imperial geographical taxonomies was facilitated by the differentiation of territories by their function within a growing global division of labor. This was increasingly thought of and justified in terms of the climatic and other environmental differences between places. Natural histories written in the nineteenth century portrayed the world increasingly in terms of North versus South than in terms of East versus West. Previously, Africa had not figured much in the emerging geopolitical imagination. This was partly because it was so unknown and thought of as basically inhospitable to Europeans. It had been something of an empty space rather than fundamental to European mapping of the world. But as a shift from civilizational judgment to scientific argument took place in the nineteenth century, Africa, the Americas, Asia and Europe had all to be considered in the great global horizontal scheme of things. The fundamental assertion was made that tropical climates encouraged idleness whereas temperate climates (such as those of Europe!) stimulated industry and innovation. This naturalization of differences in economic activity failed to note, of course, how much manufacturing and other 'advanced' activities were prevalent in climatic zones other than the temperate until killed off by European mass production and the subordination of tropical markets by temperate ones, owing little or nothing to any intrinsic character of those zones.

The attraction of this characterization to Europeans, however, is illustrated by the extent to which even late eighteenth-century British accounts of both economic development and cultural differences worked around differences, if yet crude, between frigid, temperate, and torrid zones (particularly the latter two). This sometimes took powerfully gendered forms when climate was associated with differing incidences of sexual desire and the varying status of women was tied into the fundamental civilizational differences between the different zones. To the Scottish writer Adam Ferguson (writing in 1767), for example, hot climates were zones of unrestrained passion while temperate climates produced greater control. More particularly, European women were seen by many writers as enjoying privileges denied their sisters elsewhere. This both exalted the position of European women over women elsewhere and put differences between places in sexual behavior and a wide range of associated cultural practices into the category of natural phenomena. Women in different postures and states of dress/undress came to symbolize the various regions of the world. If a crowned Britannia (in a British eighteenth-century publication) stood in for Europe, surrounded by attendants America, the East and Africa (Plate 1.2), Africa was often a naked woman and the East a beautifully draped courtesan surrounded by admiring men (Nussbaum 1995, 67–69).

Through the frequent linking of luxuries with the feminine in eighteenth-century writing, the naturalization of difference by climatic zone also associated the tropical realms from which the bulk of the new luxuries came with the feminine. This was the global counterpart to the pervasive fear of the feminization of culture and politics among European political and intellectual élites of the eighteenth century:

Plate 1.2 The crowned Britannia, a frontispiece by G. Child in *A New Collection of Voyages and Travels Vol. I* (London 1745–47). The crowned Britannia, representing Europe, is seated holding her sceptre, and flanked by her attendants America, the East and Africa
Source: Nussbaum, 1995

> As wealth came more and more from the colonies and the stock market, and as cultivation at home became more efficient, the defenders of a ruralist (and masculinist) political ideal became more and more defensive. Power was seen as moving from country to city, and what came back to the country was the iconography of surplus wealth: parks and mansions occupied for half the year by stock-market millionaires, feminized personalities. . . . The increased concentration of surplus wealth in the hands of an aristocracy and burgeoning middle class was seen as replacing an economy of need (masculinist subsistence) by an economy of desire (feminized superfluity).
>
> (Simpson 1996, 101)

The naturalization of the binary division of the world reflected in these oppositions reached its zenith in the geopolitical schemas proposed in the late nineteenth and early twentieth centuries. These played off the opposition between sea powers and land powers first noted by the ancient Greek writer Thucydides in his *The Peloponnesian War*. They expanded the geographical scope from the case of Athens versus Sparta to that of maritime trading empires (such as Britain) versus autarkic land powers (such as Germany and Russia). A European classical education finally paid off in contemporary relevance. Geographical location in relation to the oceans and land masses suggested a kind of geographical 'fate' against which agency was largely futile, save to point out the possibilities or options available to different parties depending on their relative location (Mackinder's formal geopolitical model of 1904: Figure 1.3). Historical analogy played an important part in making a model such as this 'common sensical;' putting it beyond critical scrutiny. For example, his reference to the history of barbarian invasions of Europe from Asia gave credibility to Mackinder's association of the Central Asian 'Heartland' with threat and danger. There had been danger from there in the past. The geopolitical destiny this implied provided a 'binding in time' or 'a futureness to the past and a pastness to the future that is fundamentally reassuring' (Crocker 1977, 37–39). The geopolitical models (and similar general theories reducing local differences to global physical and environmental contrasts) therefore helped to stabilize the sense of a shrinking world in which rapid change was the order of the day. The whole world had been incorporated into the European-based state system (including the territorial empires of the day). Even 'outsider' states such as Japan and the United States now aspired to follow the imperial path laid down by the Europeans. The late nineteenth century also saw dramatic shifts in space–time organization with the spread of railroads, telephones, steamship lines, foreign reporting, photography and cinema, and trade and investment flows. In a world of economic and social upheaval geopolitical models offered an appealing stability and certainty.

They also offered a view of the world that could claim grounding in the natural features of the earth. They did not express 'mere opinion.' They told it as it was. That

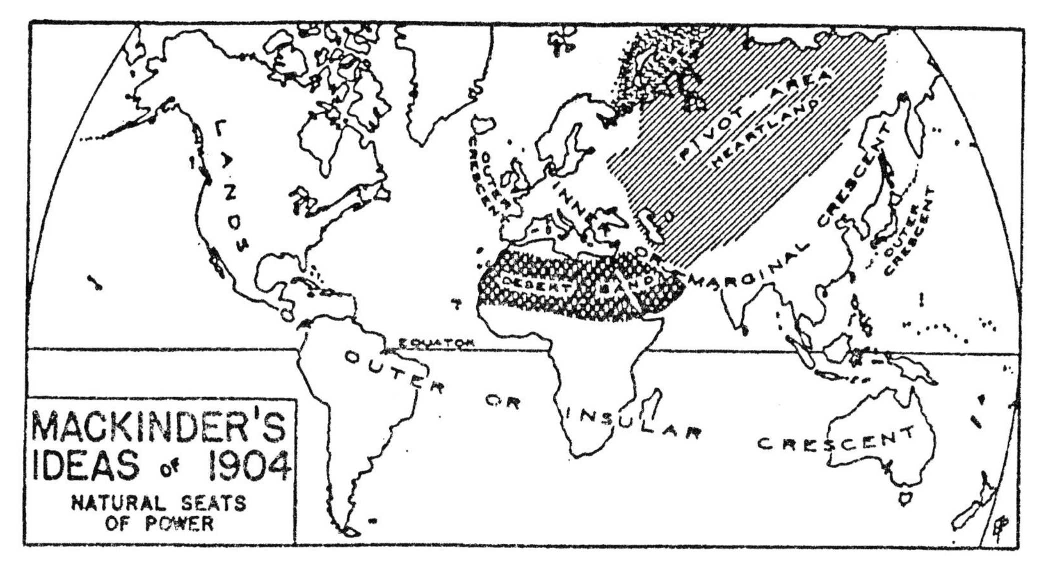

Figure 1.3 Mackinder's geopolitical model (1904). Halford Mackinder (1861–1947) was a major advocate of geography as 'an aid to statecraft,' as well as a Conservative MP for a number of years. He revised his geopolitical model on a number of occasions, although its basic geographical structure remained the same.

they did so in the context of advising different states of their global prospects did not initially appear as contradictory. Only when a revisionist Great Power, such as Nazi Germany, appeared to have succumbed to the demonic influence of a spate of geopolitical model building did the idea of a formal geopolitical model beyond the interests of a particular state finally seem problematic. Not surprisingly, therefore, by dint of their association with the crimes of the Nazi regime in Germany, formal geopolitical models tended to lose favor after the Second World War.

The American containment and domino-effect doctrines of the Cold War (referred to in the Introduction in relation to the Vietnam War) had echoes of the older geopolitical models, such as Mackinder's 'Heartland' model, even though they lacked the formal grounding in the Big Picture that these models offered. The visual metaphor of the domino effect proved particularly compelling. As the historian Frank Ninkovich (1994, xvi) has convincingly shown, from its origins in Woodrow Wilson's argument for US entry into the First World War down to the Vietnam War and beyond, the 'figurative image of toppling dominoes' tied the United States into a 'tightly linked system' which was not 'self-regulating' but required the application of force to prevent the 'chain reaction' from reaching the 'last domino:' the United States. Threats from a despotic and dangerous East (represented by the Soviet Union and 'Red' China) could only be contained by defending the first dominoes; otherwise all would be lost, as one country after another fell to the foe in a process of contagious diffusion. In another version, the process of contagion was held to be analogous to the impact of one rotten piece of fruit on surrounding ones. A Manichean (total difference in religious status between entities) religious opposition sometimes entered into the domino calculus. Hell or Satan was seen to reside in the East (as it/he had resided in the medieval 'underworld' described so imaginatively by the great medieval Florentine poet Dante). The satanic other or 'evil empire,' to use the American President Ronald Reagan's characterization of the Soviet Union, was the uniquely wicked source of all travails facing humankind, working from the weakest towards the strongest links in the armor of the Righteous rather than confronting the strongest outright. Soviet leaders used similar representations (relying more on metaphors of organic growth and decline akin to the 'rotting fruit' version of the domino theory rather than on religious imagery) in their accounts of the great global struggle between an embattled socialist experiment and an aggressive but decadent world capitalism.

By the mid-twentieth century, however, a number of dominant spatial practices and representations had underwritten the continuing connection between local differences and the binary division of the world, largely independent of formal geopolitical models. One was the growing interconnection between parts of the world as colonial economies generated regional specializations (such as cotton in the US South and rubber plantations in 'British' Malaya and the 'Dutch' East Indies) which worked as components of a multilateral trading system joining together Britain, other industrializing countries and the colonies in a global system of production and exchange. The colonized South was assimilated into the place of the ancient East; but with much the same stereotypes justifying its subordination. In the years after the Second World War the Mother Country–Colony distinction that had reproduced the East–West division for so long under colonial conditions broke down under the onslaught of decolonization movements and the emergence of the Cold War conflict between the United States and the Soviet Union. This gave way in the 1950s to the 'Three Worlds' concept of a normal, natural West (the First World) challenged by an unnatural, state-dominated East (the Second World) with each vying with the other to produce political–economic disciples in a Third World of traditional yet developing countries that became the typical representation of the enduring geopolitical division between East and West. The governments of the United States and the Soviet Union represented themselves and one another in ideal–typical terms as global agents engaged in a struggle over the heritage of the West. The Soviet Union proved especially vulnerable to the charge of oriental despotism, even though its political economy and system of governance had good Western credentials. It became the wicked Other against which the identity of United States was defined. Much domestic as well as foreign policy in the United States was made in response to perceptions of the threat that the Soviet Union posed to the American view of world order and vice versa. What is most important is that a binary geography of East versus West again dominated the understanding of how the world works. The category of the 'Third World' derived from the structural opposition drawn between the other two Worlds: an American-allied First World confronting a Soviet-dominated Second World. The regions and countries within the vast zone of the Third World, therefore, were reduced to the role they performed in the conflict between the other two rather than being meaningful in and of themselves (see Chapters 2 and 5).

The militarized nature of the conflict between the United States and the Soviet Union served to further bifurcate global space. At one and the same time, the advent of the capacity to deliver nuclear weapons over great distances almost instantaneously both devalued the military importance of territorial space through a new emphasis on 'virtuality' and yet reinforced the sense of being targeted because of where you happened to live. The net outcome was an enhanced sense of belonging to a 'targeted space,' which in Cold War terms meant being either *of* the East or *of* the West. You had to be on one 'side' or the other, given the targeting patterns of the weapon systems. Even as deterrence based on 'open' knowledge of the other side's destructive capabilities was replaced by the dissimulation implicit in decoy, STEALTH, and false radar signature technologies, 'home' still depended on which side you were on. For example, 'Duck and Cover,' the leitmotif of American civil defense programs in the 1950s (imagine school children scrambling under desks), created the illusion that it was possible to survive a nuclear attack and live to perpetuate the values of the sanctuary, the American home, that such an attack would necessarily violate.

Since the end of the Cold War in the early 1990s the Three Worlds concept has lost much of its appeal. In addition, growing disparities in economic development within the Third World category have made its use alone without reference to the other two increasingly problematic. The disappearance of an immediate nuclear threat has likewise removed that element in the East–West opposition that rested on making a totalistic military distinction between the First and Second Worlds. The emerging binary divisions seem to be either those drawn between the information 'rich' and 'poor,' relative to access to global telecommunication networks and flows of information (largely in English) about finance and production or those positing a continuing cultural East–West divide with respect to 'civilizational' conflicts emerging along global cultural fault lines (particularly that of the 'West versus the rest'). The first of these suggests a much patchier and more localized pattern of global difference whereas the latter is indicative of a continuing global division to which local differences must be assimilated. The struggle between these two visions and the degree to which spatial practices are organized around each of them will be important (along with attempts at reinstating state-centeredness) in determining what, if any, prospects the modern geopolitical imagination will have for reconstruction in the future. I will return briefly to these two competing visions at the end of Chapter 5.

CONCLUSION

World politics has developed to its present form because of the way in which the 'world' itself has been brought into focus over the past five hundred years. In brief compass, I have shown how two aspects of global visualization have produced the major geopolitical taxonomies around which world politics is organized. Seeing the world as a structured 'whole' required reaching back to older cosmographies to find a relevant model for understanding the world which Europeans started to encounter from the late fifteenth to the early sixteenth centuries onwards. That of Ptolemy provided a particularly auspicious beginning because of its openness to discovery and 'filling in' of new information. The rapidly emerging linear perspective associated with the architectural and cartographic practices of the Renaissance added a critical element: the separation of the observer from an object under vision and the object's location within a visual field which would determine the meaning that it would take. This had the effect of allowing the world to appear as a meaningful whole in which any part of it could be understood only in relation to the whole. It also led to an emphasis on the self-evident reliability of information as it appeared on maps of the whole. Maps were guides to territory, not templates framing the selection, naming, and ordering of places. These conceptual innovations provided the backdrop for a more specific and recurring tendency to divide the world into two opposing zones, each of which defined the other by communicating what it is not. The most important of these has been a persisting West–East divide (subsuming a North–South divide). Though reproduced in a number of manifestations, depending on the current mix of material conditions and related representations (see Chapter 5), this division has ancient roots in Western thought. But it also works with and confirms other basic oppositions between temperate and torrid climates, pluralist West and despotic East, sea powers and land powers, and civilized and barbarian, that relate ideas about the nature of economic and political practices to geopolitical visions of how the world works as a whole. Such essential oppositions define the horizons of the modern geopolitical imagination as it has developed from the late fifteenth and sixteenth centuries to the present. The most typical of these within the modern geopolitical imagination, that of primitive or backward versus modern, is the subject of Chapter 2.

RECOMMENDED READING

Crone, G. R. 1978 *Maps and Their Makers: An Introduction to the History of Cartography*. Hamden CT: Archon.

Gregory, D. 1994 *Geographical Imaginations*. Oxford: Blackwell.

Harley, J. B. 1989 Deconstructing the map. *Cartographica*, 26: 1–20.

Lestringant, F. 1994 *Mapping the Renaissance World: The Geographical Imagination in the Age of Discovery*. Berkeley and Los Angeles: University of California Press.

Mattelart, A. 1996 *The Invention of Communication*. Minneapolis: University of Minnesota Press.

Nussbaum, F. 1995 *Torrid Zones: Maternity, Sexuality, and Empire in Eighteenth-Century English Narratives*. Baltimore: Johns Hopkins University Press.

Philip's 1996 *Atlas of Exploration*. London: Philip's in association with the Royal Geographical Society

Ryan, S. 1996 *The Cartographic Eye: How Explorers Saw Australia*. Cambridge: Cambridge University Press.

2
TURNING TIME INTO SPACE

•

In this chapter I want to suggest that a second element of the modern geopolitical imagination has involved translating time into space. In other words, the 'blocks' of space identified in Chapter 1 have been understood in terms of the essential attributes of different time periods relative to the idealized historical experience of one of the blocks: the West. Hence, territories are named as 'primitive' vs. 'advanced' or 'backward' vs. 'modern' in relation to an idealized version of European experience.

Numerous critiques of social science thinking about social change have noted this move. My purpose in this chapter is to extend the critique in two ways. First, to argue that converting time into space, so to speak, is a fundamental characteristic of modern geopolitical thought. Second, to claim that this conception of space pervades contemporary writing about 'national development' irrespective of where in the world the state in question is located. Even though this way of thinking developed along with European colonialism, it cannot be reduced to it, in the sense of applied only to those parts of the world on the receiving end of European empire-building. Rather, the whole world has had to fit into this conception of social change, not only those parts subject to direct European economic and administrative control outside of the European realm, but including Europe itself. Not surprisingly, therefore, the mode of understanding has continued to survive long after formal colonialism has officially ended. Imperialism, in the sense of the understandings and practices of the modern geopolitical imagination stimulated by early encounters outside of Europe (and its extensions, such as the United States), lingers on even after the demise of colonialism. To illustrate the general argument about the pervasiveness of the expression of time in spatial terms I use the examples of Australia and Italy. These show that the division of space in terms of time sequences is not just a simple global one, as illustrated perhaps by the Australian example, but has become both more abstract and generic in its applications, as illustrated by the way Italy is conceived of in much contemporary thinking.

Why does this matter? For one thing, understanding of both time and space has suffered as a consequence of expressing one in terms of the other. In particular, time has lost its dynamic character as it has been reduced to a twofold categorization of space. As a result, space has also been oversimplified and reduced to a set of simple containers for either/or statements about its characteristics. The language of modern thought is filled with the fusion of time with space into binary distinctions between those areas 'ahead' or 'advanced' and those areas 'behind' or 'backward.' Indeed, one of the main organizing devices for thinking about the world, the intellectual division of labor between specialists in the 'Three Worlds of development,' students of the advanced capitalist world, students of the (former) Communist Second World, and students of the barriers to 'development' in the Third World of former colonized peoples, derives from the imposition of an idealized version of the past development of the First World onto the spaces occupied by the two other Worlds.

The juxtaposing of the temporal stage with the spatial category is vital to the modern geopolitical imagination because it provides a natural link between the European past, on the one hand, and the global present outside of the modern world, on the other, in terms of what the latter lacks and what the former has to offer to make up for this deficiency. The projection of temporal qualities drawn from a rendering of a specific historical experience onto terrestrial space, enables three political–intellectual positions. One is the tendency to *essentialize* places, or identify one trait as characterizing a particular spatial unit (e.g. caste in India, Mafia or political instability in Italy). A second is the temptation to *exoticize*, or focus on differences as the single criterion for comparison between areas. Similarities or global conundra are thereby eliminated from consideration (e.g. difficulties of social mobility everywhere, general barriers to political participation). The third is the tendency to *totalize* the comparison, or turn relative differences into absolute ones. The 'whole' of a society is thereby made entirely recognizable in any one of its parts. The whole 'block' of space is suffused with the character that is defined by the social totality (usually a 'culture' understood as a spatially indivisible unity). These have always been questionable modes of thinking. In the contemporary world of global flows of people, goods, and money they are all problematic.

The problem under consideration is not just a theoretical one: the relative adequacy of a particular account of how the world is divided up and characterized so as to make the best sense of it is also a normative problem. Places and people are not understood on their own terms but only in so far as they slot into the global scheme of things as spaces that are at this or that 'stage' of development relative to an idealized European past. The particularities and peculiarities of Europe and elsewhere are thus swept into global categories based solely on Europe's presumed more 'advanced' status. But this lays the groundwork for claiming the superiority of some places over others. This is what turning time into space does for the modern geopolitical imagination.

AUSTRALIAN JOURNEYS

Of course, the experience of European encounter with the rest of the world was always more complicated and less uniform than geopolitical categorization would lead one to suspect. The movement outwards from heart to edge of empire was uneven and produced diverse forms of colonialism. In circumstances such as those in India where outsiders confronted a densely populated and recognizably sophisticated civilization, the existing social order was largely maintained but reoriented to the economic interests of the imperial core. The colonization of Australia, however, was premised on the view that its indigenous society was primitive and its view of property conflicted with the needs of ordered settlement. This did not mean that all explorers and early settlers thought of the Australian interior as an unoccupied land freely available to all settlers. There were frequent disputes over Aboriginal land rights. There was a sense of wonder at the ability of Aborigines to survive and, to a degree, prosper in environmental conditions that Europeans saw as inhospitable. The outcome of imperial domination, therefore, was contested and contingent. But the stories told to later generations of 'new' Australians conformed much more to the heroic narrative of the 'modern' sweeping away the 'backward.' Explorers also tended to use similar heroic terms with which to characterize their own activities in confronting unknown and dangerous peoples. In every case, therefore, the explorers of Australia already knew how to interpret what they would find before they left the coast for the interior. The modern geopolitical imagination owes much of its appeal to such simplification of history: sorting empirical information into already operative sets of pigeon-holes.

The political and moral consequences of the spatial representations and practices involved in turning time into space are thus well illustrated by nineteenth-century European encounters with Australian Aboriginal peoples. To Europeans, the boundary between them and those they encountered was a line of proprietary appropriation, defining the limits of the modern in contact with the traditional. To the Aborigines, however, boundaries were 'debatable places,' which they regarded as zones for inter-tribal communication. Of course, Australia was finally 'settled.' It became a settler colony with a largely non-indigenous population. As a result, the boundaries that emerged served not to demarcate a cultural encounter but to impose a European presence. This violent imposition,

> which substituted for conversation, [was] institutionalized in the form of what is represented as 'Australia' just as other names and boundaries on the dominant, geopolitical world map are rigidified and thus removed from the possibility of encounter. . . . It is time to unread the old map and begin the process of writing a new one, a process without limit.
>
> (Shapiro 1995, 40–41)

Australia can be described as a settler colony which managed to achieve the transfer of the imperial power that had given it its first existence as a geographical entity from its core in Britain to the colony itself. Consequently, Australia does not fit the category of the Third World in which imperial power remained largely alienated from its creations; but neither does it neatly fit into that of the European First World. Rather like the popular history of the United States for Americans, the history of Australia's exploration and settlement still bulks large in the self-images of Australians and in contemporary Australian politics. Recent political disputes over finally casting off the last ties with Britain by becoming a republic and court cases over Aboriginal land claims, are reminders of the origins of the country as a settler colony (see Figure 2.1). Australian experience suggests that the terms of engagement established at the time of original exploration and settlement are not easily transcended. The nature of European engagement with a world outside of its own lives on today, therefore, in the imaginations and politics of many who had nothing to do with the original encounter.

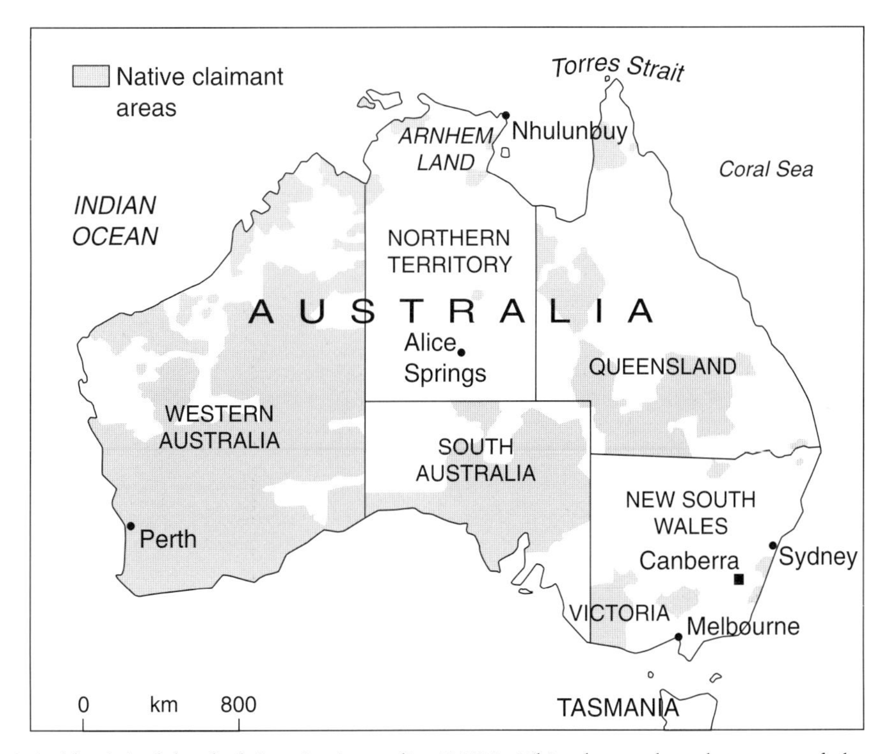

Figure 2.1 Aboriginal land claims in Australia (1997). This shows that the terms of the original exploration and first settlement of Australia continue to have reverberations well over a century later.

Australia was 'opened up' to European exploitation and settlement by large numbers of adventurers, traders, missionaries, criminals on the run and more conventional 'explorers.' There was considerable range of opinion about the native inhabitants, labeled as Aborigines from an early point in the process of exploration. What seems clear, however, is that the terms of debate were already set before any exploration took place. Two strategies of understanding predominated in the accounts of nineteenth-explorers of Australia. One was to interpret native customs with respect to either classical European culture (the ancient Greeks and Romans) or relative to stereotyped views of the East (or Orient) that Europeans thought that they knew. The explorer Mitchell (1839), for example, interpreted the Aboriginal use of green boughs of leaves upon meeting visitors as indicative of benign intentions. This involved explicit reference to the ancient Greek association of green boughs with peaceful intent. Plausible as this might be, it required the picturing of Aborigines with respect to the European past rather than to a different culture co-existing with them in time and space. On many occasions explorers compare what they think the Aborigines are up to with what they already know as exotic from the Arab, Chinese and Indian realms. This creation of the Aborigines as familiar, however, did not extend to regarding them in any way as equals of Europeans. A second strategy of understanding uses a number of devices to set the Aborigines apart as totally different from the explorers themselves. In the first place, a pre-existing language of race was used to classify the Aborigines in relation to a range of oppositions: White vs. Black, civilized vs. savage, Christian vs. heathen, rational vs. instinctive, etc. The civilized vs. savage opposition was particularly widely used to explain reactions to the explorers' presence. Thus, innate 'savage' urges were identified as explaining why Aborigines might object to the draining of water resources by a visiting explorer. That this kind of reaction was a rational response to the depredations visited by the intruder never seems to have occurred to the explorers. At the same time, the Aborigines were identified with a 'wilderness' in which rational definitions of individual ownership did not apply. This was put down to the fact that Aboriginal society was at the bottom of the global social hierarchy. This hierarchy associated peoples in the European past, such as people of the Bible, and the Orientals, against which European civilization had became defined, with the Aborigines. A pre-existing mindset brought into the field was used to interpret what was discovered.

These understandings mattered because explorers were in the vanguard of European settlement and provided a set of founding heroes for the emerging colony–nation of Australia. They recapitulated understandings that Europeans took with them wherever they went. Not that all said or thought exactly the same, more that there was a common semantic field of understanding using the same vocabulary and eliciting similar comparisons with the European past and the seemingly more familiar Orient (as described in Chapter 1). One explorer, George Grey (1841) (quoted in Simon Ryan 1996, 158), captured both this historical equivalence and the sense of European superiority in relation to the present condition of the Aborigines when he wrote in his journal about Aborigines watching his camp from the distance:

> with the help of my telescope I once distinguished their dusky forms moving about in the bush. A large flight of cockatoos, which lay between us and them, were kept in a constant state of screaming anxiety from the movement of one or the other party, and at last found their position so unpleasant, that they evacuated it, and flew off to some more quiet roosting place. Their departure, however, was a serious loss to us, as they played somewhat the same part that the geese did in the [Roman] Capitol; for whenever our sable neighbours [Aborigines] made the slightest movement, the watchful sentinels of the cockatoos instantly detected it, and by stretching out their crests, screaming . . . gave us a faithful intimation of every motion.

Not only does this description call up a classical reference familiar to British schoolboys of that era, it also implies that the Aborigines are the transgressors, in danger of trespassing upon the explorer's land. That it is in fact the Aborigines who stand in literal succession to the Romans in this historical analogy is, of course, totally lost on the observer.

Not simply 'out there,' Australia was created in the encounters of Europeans with the continent and its inhabitants. But the discovery was not just a naïve response to what was encountered. As elsewhere in the world colonized by Europeans, the encounter was actively mediated by existing ideas about the range of societies and their relative power and sophistication. From this point of view, Australia was already understood before it was discovered! And, it was understood largely in terms of the 'stages' of a European past imposed on everywhere else.

STRANGE ITALY

The conceptual grid figuring spatial differences in temporal terms has become so universalized that even differences within Europe have also come to be thought of in the vocabulary of backwardness and modernity. So, even while on a global scale Italy is modern, now the world's fifth largest economy with a per capita income bigger than that of Britain, within Europe it is often still seen as exhibiting clear signs of backwardness: political corruption and instability, organized crime, commitment to emotion more than reason, flair for decoration and presentation. As with the Australian case, a moral judgment is implied by this usage. The differences between it and its neighbors are totalized. The position seems to be that there is something lacking about Italy in comparison with its neighbors that accounts for these differences, *ergo* it has lagged behind them on the path to modernity.

How common is the invocation of the backward–modern metaphor in relation to Italy? The answer is: very common indeed. Discussion of Italy's relation to modernity and modernization has been central to debate in studies of Italy by Italians and others. Adoption of the vocabulary of backward and modern is not restricted to any particular school or grouping but is common across the political spectrum. For example, the debate over the nature of Italian fascism (1922–1943) is dominated by claims and counter-claims about its 'modern' and 'traditional' qualities, irrespective of the political commitments of the authors. All leading genres of writing about recent Italian history are also dominated by this way of thinking. I want to focus here on two examples: one drawn from English writing about Italy (Paul Ginsborg's book (1990) on the history of postwar Italy, 1943–88) and one from the work of a famous Italian anthropologist (Carlo Tullio-Altan).

Ginsborg's detailed account of Italian history from 1943 to 1988 works around a series of oppositions through which movement from backwardness to modernity is interpreted. In order of degree of abstraction these are weak state vs. strong society, family orientation vs. collective action, class and regional dualisms: workers vs. bourgeoisie, north vs. south, corrective vs. structural reform, and militance vs. *riflusso* (retreat into private life). Ginsborg operates from a standpoint sympathetic to the condition of peasants and workers in the early postwar period. He sees the period of 1943–48 as one in which opportunities to redistribute wealth and power were lost, unlike in Britain where important changes took place. Subsequent history is read in terms of the consequences of this fateful failure. But despite this failure, by the 1980s a nationalization of values had finally captured even the most traditional of Italians, though Ginsborg seems doubtful as to its permanence. This is because Italy's modernity is uniquely fragile, threatened by the persisting possibility of reversal of the current balance of oppositions. The path to modernity taken after 1948 is still subject to unpredictable shifts and pathological turns. Consequently, as Ginsborg concludes for Italy in the 1980s, Italy remains a country with 'a deformed relationship between citizen and state' (1990, 421), where 'Neither from civil society nor from the state has there emerged a new and less destructive formulation of the relationship between family and collectivity' (ibid., 418) and, more particularly, 'the lack of *fede pubblica* (civic trust) continues to bedevil southern society' (ibid., 417). Thus, for Italy, unlike Britain, true modernity is always around the corner, promised but never finally arriving.

From another perspective, however, it is not so much the absence of modernity as the continuing presence of backwardness that exercises attention. In his book, Tullio-Altan (1986) sees Italy as condemned to perpetual repetition of a condition of civic immaturity. He attributes this to the temporal persistence and geographical spread within Italy (from south to north) of the syndrome of amoral familism first diagnosed by the American political scientist Edward Banfield in his book *The Moral Basis of a Backward Society* (1958).

To Tullio-Altan, the roots of Italy's backwardness lie in the initial failure at the time of unification in the period 1860–70 to overcome the national dualism between a developing north and a backward south. He makes much of the prophesy of Mazzini (one of the leading figures in the movement for Italian unification) that 'Italy will be whatever the Mezzogiorno [southern Italy] will be' (1986, 16: my translation). Two phenomena are identified as largely responsible for the 'southernization' of Italian society: *clientelismo* (the exchange of votes for favors) and the practice of *trasformismo* (collaboration among politicians by the exchange of favors). The spread of these practices has prevented the development of a national ruling class with a truly national orientation. By its continuing geographical fragmentation Italy has been trapped in a condition of cultural backwardness compared to more successful national societies.

The main claim is that the Italian 'political class' embodies both a culture that is particularistic and a parallel political practice that is based on the exchange of favors. The presumption is that these are a straightforward inheritance from the past. But what if they are new rather than old? What if they are elements of a 'modern and effective system of power designed to integrate into the national society masses of people who are dangerously inclined to claim democratic participation and their own emancipation,' in other words, 'a modern political culture' (Signorelli 1986, 45)? Furthermore, the ideal of civic maturity is hardly realized anywhere on earth so why hold Italy to such an impossible standard? Discouraging political participation and subverting political institutions for personal gain are features of most political societies. They are definitely not unique to Italy.

The attribution of backwardness only makes sense, therefore, if the 'backward' features of Italy are implicitly compared to some ideal of modernity. For Tullio-Altan these would obviously include the absence of the vote–exchange nexus that he gives so much attention to in his account of Italian backwardness. But he also offers some clues as to the more positive features of modernity. He draws a stark opposition between societies organized with reference to a social conscience based on individual responsibility, which he sees as produced by the Calvinist branch of the Protestant Reformation, and those, such as Italy, in which an 'amoral familism' prevails. This is Banfield's idea that a modern conscience fails to develop when people's moral horizons are defined by the bounds of a nuclear family. Italy's backwardness, therefore, derives ultimately from the failure to have experienced the Protestant Reformation.

In fact, the Italian case indicates something of the arbitrariness attached to attributions of backwardness and bolsters the suspicion that they may be political in origin. For example, there is much evidence that Italy has long had a more compact and

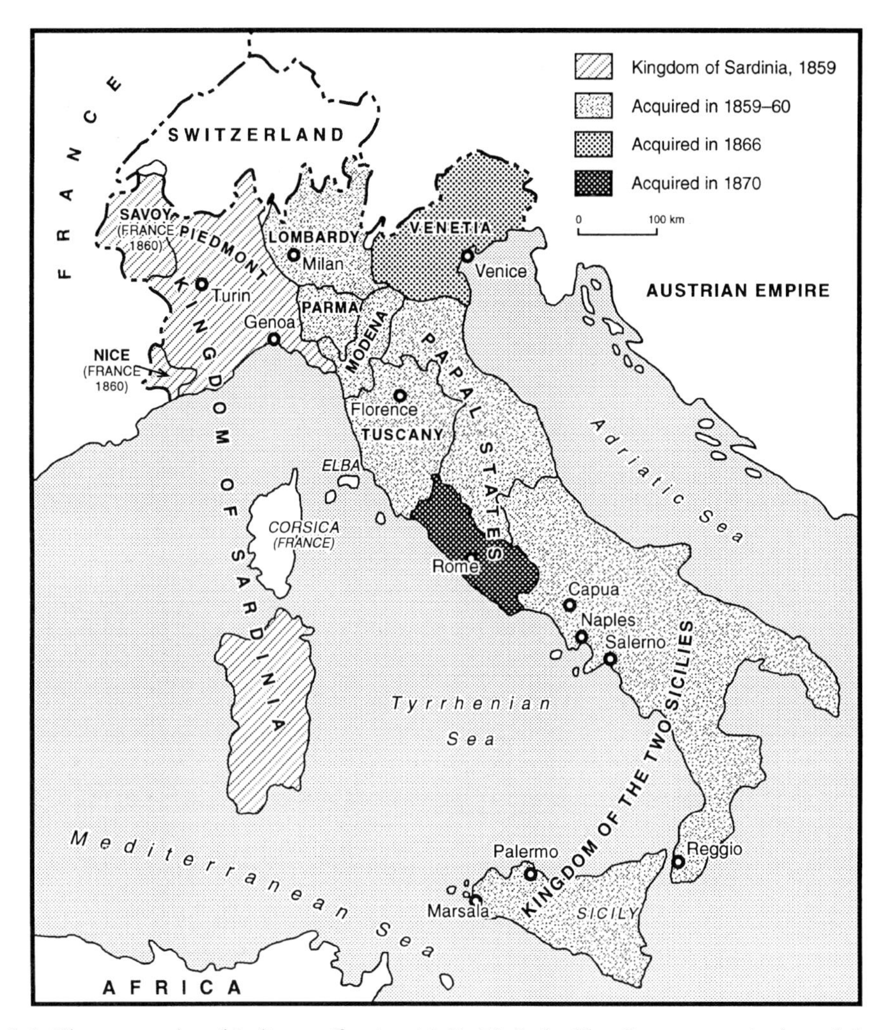

Figure 2.2 The geography of Italian unification 1860–70. Italy, like Germany, united much later than the other large European states. Large areas were controlled by Austria and the peninsula as a whole had long been politically fragmented

nationally orientated political élite than many other European nation–states, including France and Germany. Also, if one focused on technology, industrialization or consumerism as attributes of modernity, then Italy could hardly be construed as backward. The imprecision of the terms 'backward' and 'modern' gives them a mysterious quality that allows them to be used to paint a picture using whatever materials are suitable to make the particular case. In this case the purpose seems to be identifying the south as the source of Italy's recent political woes and labeling as indicators of backwardness those features of Italian politics and society that are most strongly associated with that region.

Given this arbitrariness, the question inevitably arises: why should attributions of backwardness and modernity have such appeal? In the Italian case two factors stand out. The first is that there is a widespread sense on the part of Italian intellectuals of the failure of Italy to 'mature' as a political entity. This is not an imposition by outsiders as an attractively simple innocent insiders vs. evil outsiders account might have it. This view has deep roots in Italy itself. For example, the main theme of the great Florentine writer Machiavelli (1469–1527) in such Renaissance-era works as *The Prince* and *The History of Florence* was the civic corruption of his epoch compared to the civic excellence of the ancient (republican) Romans. The power of the Popes and intervention in Italy from 'outside' were indicted as the main culprits. Again, the main figures in achieving Italian unification in the nineteenth century blamed the reactionary Austrian and Bourbon (Spanish) regimes that governed large parts of the peninsula and islands for the social and economic decline that Italy experienced after the Renaissance (Figure 2.2).

In the second place, Italy has also been plagued ever since unification by a disunity based on geo-economic and linguistic fragmentation. The city–states so closely associated with the European Renaissance (Florence, Venice, etc.) failed to produce an integrated state that would have led Italy towards the model followed by France, Spain and England. The variety and density of cities with their own hinterlands worked against the ready creation of a territorial state with a dominant capital city. Foreign rulers found in Italy propitious circumstances for a strategy of divide-and-rule. The continuing secular power of the Roman (but Catholic) Church has continually frustrated attempts at creating a national state in its homeland. As a result, Italians have had to look outside of Italy for inspiration, to the extent that 'admiration for foreigners has been the principal ingredient of Italian nationalism' (Lanaro 1989, 212: my translation)!

SPACES OF BACKWARDNESS

One consequence of Columbus's return from his famous voyage of 1492 was a heightened sense among European intellectuals of a hierarchy of human societies from primitive to modern. It is surely no coincidence that in conventional accounts of the periods of history, 'modern' history begins with the era of Columbus. However, the simple juxtaposition of new worlds against a familiar 'old world' from which the discoverers came is an altogether too simplistic view of what happened. All societies define geographical boundaries between themselves and others. Sometimes the world beyond the horizon is threatening, sometimes it is enticing. But not all societies engage in portrayals of the other societies they come across as 'primitive' or 'backward.'

In fact little is known of how exactly early modern Europeans assimilated 'exotic peoples' into their understandings of historical geography. In a major review of scholarship covering the sixteenth and seventeenth centuries, however, Michael Ryan suggests that the major means was through the assimilation of the exotic into Europe's own pagan and savage past. 'In the triangular relationship among Europe, its own pagan past, and the exotic, the principal linkage was between Europe and antiquity' (Ryan 1981, 437). The categories of 'pagan' and 'barbarian,' discovered as an inheritance from the European Ancients, were deployed to differentiate the newly discovered worlds from the old one. Thus, a conception of the temporal transition through which the European social order had been transformed was imposed upon the spatial relationship between the new worlds and Europe in its entirety. The religious dimension was especially important in reading the new pagan worlds as standing in a relation to the (European) Christian world as that world stood in relation to its own pagan past.

This is not entirely surprising if it is remembered that the first 'discovery' of the new geographical worlds coincided with the rediscovery of Europe's own ancient past in the European Renaissance (noted in Chapter 1). Indeed, as Mandrou (1978, 17) recounts:

> The new worlds that fascinated the intellectuals of the sixteenth century were not so much the Indies – West or even East – but those ancient worlds which the study and comparison of long-forgotten texts kept revealing as having been richer and more complex than had been supposed.

Of course, Italy was the center for this new activity, associated as it was with monastic libraries, universities, and the recovery of ancient texts. Ironically, given the challenge to ecclesiastical authority that the rediscovery of the Ancients could entail, it was from within the Church that the 'new learning' arose: 'Italian cities, richer in Churchmen than any others in Europe, and closer to the papal authority, constituted the setting that was most apt to stimulate the study of ancient texts and pre-Christian thought' (ibid., 23).

The articulation of spatial differences in temporal terms was reinforced by the 'taxonomic lore' that Renaissance-era Europeans learned from the Ancients. As Edward Said (1978, 57) has suggested, much ancient Greek drama involved demarcating an 'imaginative geography' in which Europe and Asia are rigidly separated. 'Europe is powerful and articulate; Asia is defeated and distant . . . . Rationality is undermined by Eastern excesses, those mysteriously attractive opposites to what seem to be normal values.' An image of essential difference with roots sunk deeply in the primordial past was used to invent a geography that had no real empirical points of reference. In terms of such categories as race, property, oligarchy and economy, the Orient (and non-Europe in general) was claimed as 'the negation of all that was being claimed for the West, by polemicists knowing, in fact, very little about it' (Springborg 1992, 20). From this stereotyping derive such views as the association of democracy with the West and despotism with the East, history with the West and cultural stasis with the East, ingenuity with the West and conformity with the East. These views never rested on empirical evidence but rather on *a priori* assertion. Not only did 'West' and 'East' have long histories of interaction and interchange; Christianity itself came from the East, but the histories of each were complex and multifaceted. Neither was the historic 'home' of either democracy or despotism. The reduction of cultural complexity to a simple two-region opposition had discrete historical origins (also see Chapter 1). The 'Ottoman peril' (threat of Turkish invasion) of the Renaissance period gave particular credibility to the sense on the part of vulnerable Europeans of a profound chasm between the familiar world of Europe and the exotic world of the Oriental Other. In Europe 'The Turkish threat worked toward reviving a waning loyalty to the *Respublica Christiana* [the idea of Europe as Christendom] and gave new life to the old cry for peace and unity in a Christendom subject to the pope' (Schwoebel 1967, 23). However far from Europe Europeans might venture, this deep sense of the civilized versus primitive and Christian versus heathen traveled with them. The explorers and settlers of Australia, therefore, could be expected to arrive there with prior expectations about the others they might meet. They certainly did not show up without a vivid sense of the differences that separated them from their more immediate neighbors such as Turks, Moors and Arabs.

As the European states emerged from the dynastic struggles and religious wars of the seventeenth century and embarked upon their schemes of empire building outside of Europe, comparisons of themselves to the ancient world, especially the 'model' provided by Rome, proved irresistible. Lord Lugard (1926, 618), the British ruler of northern Nigeria, was to maintain towards the end of the colonial period that Britain stood in a kind of apostolic succession of empire: 'as Roman imperialism . . . led the wild barbarians of these islands along the path of progress, so in Africa today we are re-paying the debt, and bringing to the dark places of the earth . . . the torch of culture and progress.' Perhaps the most important exponent of this Roman-inspired view of history was the German philosopher Hegel (1770–1831), especially in his *The Philosophy of Right*, first published in 1821. On the basis of the relative extent of the absolute sovereignty of the state and what he called its 'ethical substance,' the nation, Hegel divided the world into four historical realms arranged hierarchically, with the Oriental (India seems to have figured prominently in his thinking) as the lowest, the Germanic as the highest (surprise!), and the ancient Greek and Roman worlds, as the precursors of the Germanic, in between. In Hegel's understanding (which fits in so well with that of later generations of Italians), the achievement of an integrated nation–state was a necessary condition for achieving moral identity (see Chapter 3). From this point of view, Hegel is particularly important as a philosopher of the geopolitical imagination. He saw the world as a whole, dividing it up into geographic chunks on the basis of levels of political development *vis-à-vis* an interpretation of European experience of which the most important part was the European nation–state, in the process of replacing a greater variety of forms of political organization (dynastic empires, papal states, feudal territories, systems of shared power such as the Holy Roman Empire, etc.) at the time he was writing.

Beginning in the late eighteenth century the resort to classical precedent to understand spatial differences in the social order was put on a scientific foundation. Pragmatic common sense was backed up by explanation in terms of natural processes or by analogy to natural processes. It became increasingly popular to see social change as a transition from one stage or level of development to another. As the nineteenth century wore on, and in imitation of discourse within biology, this view was elaborated upon as an evolutionary movement from a lower to a higher level of organization. This trend took two directions. One direction was that 'race' shifted in meaning from the looser sense of groups differing as much civilizationally as physically to a much stronger sense of complexion and physical stature as determining civilizational differences. If an earlier generation of Europeans, including many of the intellectuals of the Enlightenment such as Kant (1724–1804), had used notions of racial difference to emphasize their break with the 'Dark Ages' in which the rest of the world was still living, by the mid-nineteenth century race had taken on, in the writings of such anti-Enlightenment figures as Gobineau and Le Bon, the connotation of *the* determining factor in explaining human history. A second direction was that some parts of the world were increasingly seen in terms of levels of development, i.e., levels of economic growth and social and political progress, that Europe had experienced previously. But this difference no longer need condemn the benighted to perpetual backwardness. If the experience of the 'developed' was imitated, then those that lagged behind could begin to catch up. The idiom of what the Indian historian Ranajit Guha (1989, 287) terms 'Improvement' came to prevail over that of 'Order.' The distinction no longer lay primarily in essential difference that could not be transcended but in the possibility of overcoming backwardness through imitation. The future of the backward lay in repeating what Europe had done.

The power of this latter idea is indicated by the degree to which political élites around the world have enthusiastically adopted the ideal of modernity as a recapitulation of European and, more recently, American experience. For example, even the apparently anti-Western governments of China since 1949 have used a rhetoric appealing to a catching up or overtaking of Europe. At the height of the Maoist period in the late

1950s, when Western ideas were supposedly anathema to the regime, the economic plan of the time, the Great Leap Forward, invoked the appeal to *chao ying gan mei* (literally 'overtaking England and catching up with the United States'). This reflected a long-standing commitment of the Chinese élites (irrespective of political ideology) to the idea of *zhong xue wei ti, xi xue wei yong* ('Chinese learning for fundamental principle, Western learning for practical use'). The problem has been that it has proved difficult, if not impossible, to keep Western influence confined to the technological realm. At the same time, it has never been entirely clear what precisely is 'purely' Chinese to defend from outside assault. The tendency has been to assert a Chinese essentialism or total difference with everywhere else based, above all, on the difficulties of material existence in China. This mandates a narrow definition of human rights associated with feeding the people, even though there is nothing about the history of China that would necessarily lead to such a limited conception. There is nothing particularly Chinese about denying freedom of speech or trial by law.

By the close of the nineteenth century, however, modernity was increasingly conceived of within Europe as the form of society in which social interaction is rationally organized and self-regulating. The German economist/social historian Max Weber's (1864–1920) theory of rationalization provided the most important account of this modernity within an evolutionary conception of social change. For Weber, writing at the end of the nineteenth century, the rationalization of social life in the modern (Western) world involved the increasing regulation of conduct by instrumental rationality; using criteria of merit in employment, for example, rather than 'traditional' norms and values using caste, family and kinship. Weber himself was less than enthusiastic about this process of modernization. He associated modernity with the rise of bureaucracy and impersonal regulation of social behavior. But many of his sociological disciples have had few doubts. The version of Weber's theory disseminated in the English-speaking world by Talcott Parsons, a leading American sociologist of the 1950s and 1960s, dissociates 'modernity' from its modern European origins and stylizes it into a spatio-temporally neutral model for processes of social development in general. It thus loses any connection to its origins as a critique of a specific social trend. In this rendering, modernity, thoroughly confused by now with American society, becomes a social model to which other 'less developed' societies can aspire.

This way of thinking was possible only because time was thought of as co-extensive with the world. This took the simple form of here (the West) was now and there (the Rest) was then. All the world was united in a single, one-way history. If technological developments such as the railway and the telephone worked to create a sense of global simultaneity, the understanding of different places in relation to the single trajectory of history worked in the opposite direction, towards an absence of 'co-evalness:' the sense of societies affecting one another at one and the same time. The idea of temporal evolution determining geographical differences won out over the idea of such differences resulting from the contemporary interplay between local societies and influences emanating from over the horizon.

A final boost to the designation of areas as backward or modern came from the ideological combat of the Cold War in which the two modern Worlds of capitalism and communism struggled for dominion in the backward or traditional Third World. Although later adopted as a symbolic referent for the solidarity of former colonized peoples, the term 'Third World' was never a particularly useful empirical designation. Its meaning was premised upon the prior existence of two competing models of development that would allow for no alternatives. This accounts for the origin of this vocabulary in the early 1950s, at the outset of the Cold War between the United States and the former Soviet Union. The backwardness of the Third World was necessary to define the modernity of the other two. In that zone the two modern Worlds could compete for converts to their respective models of political economy. In this translation of time into space, the Third World represented what the other two Worlds used to be like. Their claims to modernity rested on comparisons with those parts of the world that still had not reached their 'levels.' There could be a modern, therefore, only if there was a backward against which to compare it. Both sides in the Cold War saw their competition with one another in this way. Their pasts were mirrored in the present and future of the rest of the world.

CONCLUSION: FROM METAPHOR TO MYTH

This criticism is all very well, one might say, but surely the vocabulary of backward and modern is nothing more than an evocative metaphor that helps to communicate the differences between Australia or Italy and two ideal types of society – the backward and the modern? I would argue that the vocabulary is now much more than this; that it organizes and directs thinking about the 'nature' of such places as Australia and Italy. From this perspective, primitive Australia and backward Italy are myths, the idealization involved forgotten as the metaphor has substituted for analysis. This is not to say that the metaphor is false in all its usages, say, as a description of differences in technological prowess, only that it functions in the cases examined here, and in the modern geo-political imagination, as a fable more than as a mere communicative device.

The literary scholar Frank Kermode makes a distinction between 'myth' and 'fiction' that may be helpful here. In his view, a fiction is 'a symbolic construct ironically aware of its own fictionality, whereas myths have mistaken their symbolic worlds for literal ones and so come to naturalize their own status' (Eagleton 1991, 191). The line between the two is fuzzy rather than hard and fast, as all fictions can become myths once established and widely disseminated. This is what has happened to the metaphors of primitive Australia and backward Italy.

It is something of a conventional wisdom among European and American intellectuals that modernized societies (Europe, particularly England, and the United States are the paradigm cases on a global scale) are rational and secular to the exclusion of traditional or metaphysical myths about their founding and nature. As argued earlier in the chapter, this position has become a central element in the backward–modern metaphor itself. Modernity is, by definition, life without myth. But perhaps it is more that our most cherished European myths, such as backward vs. modern, are simply ones without hope of providing a suddenly better world in its entirety, merely naturalized fictions that give meaning to speculation about the historical trajectories of particular societies. Although it is part of their appeal, at least in the idiom of improvement, that they do hold out the hope that others can become more like us.

It is precisely the 'commonsensical' quality of the opposition between the traditional and the modern which gives it mythic power. Since Columbus first returned from his trans-Atlantic voyages we have become so used to characterizing geographical differences in idealized temporal terms that we cannot see any problems with this way of thinking. The explorers of Australia certainly thought this way. They assimilated empirical information to the oppositions (primitive vs. civilized, etc.) that they worked with. Later generations have continued to read the *whole* world through the basic opposition of backward and modern. The message of subordination implied by this pairing is sent masked as one of 'natural history.' So, even as the original context for turning time into space was extra-European, following the early voyages of discovery and conquest, the same message can now be sent inside Europe, indicating that the essential message is still one of geopolitical dominance and subordination rather than justifiable empirical claims about national development. The ultimate irony, therefore, given Columbus's own origins in an Italian city–state, is that in a European context Italy has been more readily portrayed as backward than modern. A temporal metaphor initially applied to make sense of the spatial 'gap' between the New Worlds and the Old one has become a preferred way of dealing with, for example, Italian differences relative to an idealized European modernity. In so doing, the intrinsically normative character of the terms backward and modern has been obscured. Yet it is their nature as moral judgments about people and places that gives them such power. They tell us who is in charge and why they should be. At least Columbus and his contemporaries seem to have understood that. Perhaps it is time political geography owned up to ethical complacency for ignoring the moral dimension of turning time into space? But that would require abandoning one of the key tenets of the modern geopolitical imagination.

RECOMMENDED READING

Eze, E. C. (ed.) 1997 *Race and the Enlightenment: A Reader*. Oxford: Blackwell.

Fabian, J. 1983 *Time and the Other: How Anthropology Makes Its Object*. New York: Columbia University Press.

Goody, J. 1996 *The East in the West*. Cambridge: Cambridge University Press.

Jacobs, J.M. 1996 *Edge of Empire: Postcolonialism and the City*. London: Routledge.

Mandrou, R. 1973 *From Humanism to Science, 1480–1700*. London: Penguin.

Mason, T. 1988 Italy and modernization: a montage. *History Workshop*, 25/26: 127–47.

Rattansi, A. and Westwood, S. (eds) 1994 *Racism, Modernity and Identity on the Western Front*. Cambridge: Polity Press.

Ryan, S. 1996 *The Cartographic Eye: How Explorers Saw Australia*. Cambridge: Cambridge University Press.

Said, E. W. 1978 *Orientalism: Western Conceptions of the Orient*. New York: Vintage.

3

A WORLD OF TERRITORIAL STATES

•

In the modern geopolitical imagination power has been defined as the ability to make others do something you desire and, at least from the nineteenth century onwards, it has been exclusively associated with territorial states that are usually presumed to be nation–states (ones where there is a close match between membership of a distinctive nation and the boundaries of a particular state). In this chapter I want to explore these contentions in some detail and to show how the 'spatiality' or geographical organization of power is not necessarily tied for all time and all places to the territoriality of states. The 'state-centered' account of spatiality of power, what I term the 'territorial trap,' is the historical projection of a world in which power over others is envisioned as 'pooled up' in the hands of equivalent units of territorial sovereignty, usually for the most important states militarily, the so-called Great Powers.

Most explicit in the case of political geography and the study of international relations, but common throughout the contemporary social sciences, the conventional understanding of the geography of power is underpinned by three geographical assumptions: first, that states have an exclusive power within their territories as represented by the concept of sovereignty; second, that 'domestic' and 'foreign' affairs are essentially separate realms in which different rules obtain; and finally, that the boundaries of the state define the boundaries of society such that the latter is 'contained' by the former. These assumptions reinforce one another to produce a state-centered view of power in which the space occupied by states is seen as fixed, as if for all time. Thinking about the spatiality of power is thus put beyond history by assuming an essential state-territoriality to the workings of power.

This perspective did work particularly well for the state-centered world that began to develop in the nineteenth century (see the section on Naturalized Geopolitics in Chapter 5). It made sense in the context of that time to see trajectories of economic and social change as increasingly characterized in terms of the experiences of the bits of space delimited by the geographical boundaries of states. Businesses and trade unions, representative politics, and social life were increasingly organized on a state-by-state basis. But there was also a normative element to thinking about power largely in terms of states. As a reflection of the burgeoning nationalism in Europe (and increasingly elsewhere) in the nineteenth century, politics was seen as *best* thought of in terms of national states. Ideas of distinctive 'national characters' and their reflection in military, sporting, technological, artistic and educational prowess became widely accepted.

At the same time, the new social sciences (economics, sociology, political science) used the territories of modern statehood to serve as a fixed and reliable template for their investigations into a wide range of phenomena. The 'modern' (European) world was seen as one in which local communities were in eclipse to the rising sun of 'societies' based on the nation–state. *Gesellschaft* (society) was replacing *Gemeinschaft* (community) as the dominant cultural–geographical ordering principle. In this way a largely implicit 'methodological nationalism' came to prevail in political and social thought. Currents of thought allowing for more complex views of the geographical scales at which social, economic, and political processes could take place were effectively marginalized.

The modern territorial state was underpinned, therefore, by the claim that it was the people's 'mentor' in the cult of the nation. At its extreme, for the French Revolutionary disciples of Rousseau, the nation–state provided the basis for re-establishing a religious foundation to political authority. Only now, unlike in the person of the emperor-god or divinely appointed monarch, the *educating state* would give citizens a feeling of moral unity and identification with the father (or mother) land. This 'sacralization' of the nation gave the territorial state an increasingly competitive advantage over other possible types of spatial–political organization such as confederations, loose empires or city–states.

The drawbacks of a state-centered perspective on power have only recently become obvious. On the one hand, this owes much to perceived changes in the ways in which states relate to one another and to the emergence of a global society in which states must share power with other types of actors. We live in an epoch in which the declining military viability of even the largest states, growing global markets, expanding transnational capitalism, and modes of governance alternative to that of the territorial state (such as the European Union, the various UN Agencies, the World Bank and the International Monetary Fund) have begun to undermine the possibility of seeing power as solely a spatial monopoly exercised by states.

On the other hand, the problem is more profound than one of a mere 'goodness-of-fit' to the changing economic and social conditions of the contemporary world. 'State-centricity' has finally been recognized as the main strategy of modern intellectuals of all political persuasions in limiting the definition of power to that of a coercive instrument and restricting 'politics' to the domestic realm of *the* state (a way of thinking that has its roots in the works of the ancient Greek philosopher Aristotle as well as those of the early modern Florentine political theorist Machiavelli). Representing space as state territoriality also serves to put statehood outside of time, because of the strong tendency to associate space with stasis or changelessness, and thus to impose an intellectual stability on the world that would otherwise be difficult. As a result, state-centricity has *continuing* normative attractions for both intellectuals and political activists even as the empirical reach of states to control and regulate recedes. It provides a grounded set of social–geographical units for both longitudinal and cross-sectional data analyses. It offers a set of concrete institutional opportunities (however weakened or compromised in effective performance) for political action.

THE TERRITORIAL TRAP

Three analytically distinct but invariably related assumptions underpin the territorial trap: thinking and acting as if the world was made up entirely of states exercising power over blocks of space which between them exhaust the politico-geographical form of world politics. The first, and most deeply rooted, is that modern state sovereignty requires clearly bounded territorial spaces. The modern state differs from all other types of organization by its claim to total sovereignty over its territory. Defending the *security* of its particular spatial sovereignty and the political life associated with it is the primary goal of the territorial state. Vested at one time in the person of the monarch, or other leader within a hierarchy of 'orders' from the lowest peasant to the warriors, priests, and nobles, sovereignty is now vested in territory.

The second crucial assumption is that there is a fundamental opposition between 'domestic' and 'foreign' affairs in the modern world. This rests on the view common to Western political theory that states are akin to individual persons struggling for wealth and power in a hostile world. One state's economic and political gains always come at the expense of others. Only inside the boundaries of the state, therefore, are civic culture and political debate possible. Outside, reasons of state (pursuit of the state's interests) rule supreme. This fixes processes of political and economic competition at the level of the system of states.

Third, and finally, the territorial state acts as the geographical 'container' of modern society. Social and political organization are defined in terms of this or that state. Thus, we speak and write unself-consciously of 'American' or 'Italian' society, as if the boundaries of the state are also the boundaries of whatever social or political process we might be interested in. Other geographical scales of thinking or analysis are thereby precluded. Often this is because the state is seen as the guarantor of social order in modern societies. The state substitutes for the self-reproducing cultural order that can be found in so-called traditional societies (this ties back into the theme of Chapter 2).

Together these three assumptions underpin a timeless conception of statehood as the unique font of power in the modern world. The first one dates from the period in European history when sovereignty shifted from the person of the monarch to the state and its citizens. In Europe, this did not happen overnight. It lasted from the fifteenth to the nineteenth centuries. The second two date from the past one hundred years, although the domestic vs. foreign opposition has roots in the doctrines of seventeenth-century economic mercantilism. Together they serve to put the modern territorial state beyond history in general and the history of specific states in particular. They define a world made up exclusively of similar territorial actors achieving their goals through control over blocks of space.

A CASE IN POINT: STANDARD INTERNATIONAL RELATIONS THEORIES

The importance of the territorial state and the similar roles it performs within different theories can be seen in the writings of such influential contemporary international theorists as Kenneth Waltz (1979) and Robert Keohane (1984). They are influential exponents at either end of the continuum stretching from 'realism' to 'liberalism;' the main positions on state power around which the modern geopolitical imagination has tended to operate in the twentieth century.

Waltz is concerned with what he calls the 'structures of inter-state relations,' altogether excluding the domestic character of states from explanatory consideration. To him, the structure of the international system has three features that count: it is anarchic, without higher authority; states all perform the same functions and are equivalent units; and there is an uneven distribution of resources and capacities among states. From these fundamental features he draws the following inferences: that at any moment the system's shape as a whole is determined by the number and relations between the Great Powers (the ones with most resources and capacities) and that the balance of power between the Great Powers is the key mechanism in world politics. From this point of view, therefore, from 1945 until 1990 the international system was a bipolar balance of power between two Great Powers (the United States and the Soviet Union) in contrast to, for example, the multipolarity of the early nineteenth century in Europe. What drives the system is fear of domination by others. States are thus understood as unitary actors with each state trying to maximize status relative to others (see Chapter 4 for a longer discussion of how important this has been for the modern geopolitical imagination). No entities other than states are involved, by definition, in international relations. World politics is entirely about international (i.e. inter-state) relations.

In apparent contrast, Keohane is interested in how cooperation can take place between states without a dominant Great Power. He argues that there are important incentives for cooperation between states that work against the competitive pursuit of coercive power in an anarchic world, even though he accepts the reality of this world.

Treaties, agreements and formal international institutions do restrict state conduct. This is because states agree to constraints when the benefits that derive therefrom outweigh the costs. States are seen as utility rather than status maximizers. But states remain the only significant actors in this account. The utility game is one only they play.

Despite their differences over status or utility maximizing, therefore, these theorists share the commitment to a state-centered world or, more precisely, a Great-Power-centered world. This is a conception inherited from a long line of political thinkers and practitioners. It is a vital ingredient of the modern geopolitical imagination.

THE SPATIALITY OF POWER IN THE MODERN GEOPOLITICAL IMAGINATION

COERCIVE POWER OVER BLOCKS OF SPACE

Three features of the conventional view of power and states have been crucial in limiting understanding of the spatiality or geographical organization of power entirely to states. These made the territorial or spatial block perspective on power problematic long before recent changes in the workings of the international political economy called it more fully into question. The first feature has been the definition of power implicit in the modern geopolitical imagination as the capacity to coerce others into doing your will (or power *over*). This leads to a notion of power as a monopoly of control exercised equally over all places within a given territory or geographical area by a dominant social group or élite ('despotic power'). This misses both the contingency and fragility of the 'infrastructural power' (state provision of public goods and services, etc.) upon which the legitimacy (the right to rule recognized by a population) of modern states largely rests. The sociologist Michael Mann (1984) has pointed out in some detail the critical role of infrastructural power in distinguishing modern territorial states (both bureaucratic and authoritarian) from both feudal and classic–imperial types of rule (Table 3.1). With the ability to provide centrally and territorially organized services that other organizations cannot, the territorial state is no longer entirely the creature of state élites (and their 'despotic' power). It has an autonomous source of power in its coordinating and directing roles. But this *relative* autonomy depends upon states delivering a set of services that cannot be provided in some other way. This, of course, opens up to challenge both the regime (current institutions) and the state if the state cannot be depended on to deliver the goods.

This redefinition, however helpful in pointing to the dependence of modern state power on what it does for its population, completely ignores the degree to which power is inherent in all human agency. All social practices involve the application of power; the

Table 3.1 Two dimensions of state power and the four ideal types of state they define

		Infrastructural Power	
		Low	*High*
Despotic Power	*Low*	feudal	bureaucratic
	High	imperial	authoritarian

Source: Mann 1984, 188

ability to engage in actions towards the completion of socially sanctioned goals (or power *to*). From this point of view, power is not some thing or potential vested solely in states (or associated political institutions) but the application of agency inherent in all social action to achieve chosen ends. Territorial states are one type of concentration of social power that emerged in specific historical conditions in which state territoriality was practically useful in fulfilling the objectives of both dominant *and* subordinated social groups. Today, we see the emergence of local and regional governments and supra-regional communities in the application of infrastructural power for such ends as economic development and political identity, usually without the coercive power traditionally associated with territorial states. Of course, such alternative spatial configurations of power to that of the modern territorial state are not entirely new. For example, the Hanseatic League, the Swiss Confederation, the Holy Roman Empire, the Iroquois Confederation, the Concert of Europe and the early United States are familiar examples of alternative systems of power and authority to that of the 'Westphalian' system of territorial states. The actual existence of these kinds of institutional arrangements points to the range of possible ways in which power can be organized spatially. They suggest that forms of power are generated, sustained and reproduced by historically and geographically specific social practices, rather than given for all time in one mode of spatial organization: that of state territoriality. Indeed, there was nothing inevitable about the emergence of the modern system of territorial states. Until the nineteenth century even their monopoly over coercive power was easily challenged, for example by pirates, and alternative arrangements for the geographical organization of centralized power were widespread, for example in overlapping jurisdictions such as the Holy Roman Empire.

More radically, the power of states over their populations and in relation to one another can be understood as resting on power 'from below.' In other words, the territorial state draws its power in capillary fashion from social groups and institutions rather than simply imposing itself upon them. From this point of view, power is present in all relationships among people and animals and the power of the state relies on the wide range of sources it can tap into. This can be termed a non-sovereign conception of power, in contra-distinction to that view which sees power as flowing from a single (sovereign) source, such as the state. In this construction, power is best thought of as equivalent to the energy moving a circulatory system rather than a mechanical opposition between a source of power, on the one hand, and an obedient (or truculent) subject, on the other. There are multiple points at which consent and resistance come into play in expanding and restricting the interplay between states and subjects and, hence, in defining the state's effective territoriality: how well it dominates its claimed block of space. The spatial monopoly of power exercised by a state is not and cannot be total when its power derives from that given up by and potentially retaken by others.

POWER AS COERCION IN INTERNATIONAL RELATIONS

The second feature of the association between states and power in the modern geopolitical imagination is that coercive relations between states have usually been seen as the only way in which power is exercised beyond state–territorial boundaries. Even cooperation between states, as in the Keohane example cited earlier, is interpreted as 'getting your own way' in disguise! Indeed, the practices of politics, group divisions and struggles over 'the good society' and who gets what, when, how and where, within state boundaries, are typically contrasted with reason of state (expedient action to defend or further the interests of the state itself) beyond them. Democratic political theory, for example, has been largely limited to the possibilities of political representation and participation within states and not their prospects among, between, or beyond them. At the level of the state system, the concept of 'hegemony' is often used to indicate the domination exercised by a particular state over others during a specific historical epoch. This definition is indicative of the central position given to despotic or coercive power in international relations, irrespective of the specific 'relation' in question. Yet, the concept of hegemony can be given a different meaning, closer to one originally suggested by the Italian Marxist thinker Antonio Gramsci (1891–1937), that refers to the power implicit in dominant practices that govern society, both within and beyond state–territorial boundaries. In this construction, therefore, world politics involves a variety of social practices that require the deployment of power, not simply military coercion by states. The identities and interests of states (and other actors) are formed in interaction with one another and in the nexus between global and local social practices. Hegemony refers to the nature of the dominant social practices in a given historical epoch and how they bind together the various actors into a global society. The dominant practices may benefit one state disproportionately (such as Britain in the mid-nineteenth century and the United States since the Second World War; see Chapter 5), but the costs and benefits (both economic and cultural) can be more diffusely distributed among all of the actors (both state-affiliated and otherwise) subscribing to the contemporary 'principles' of international life – such as those defining the modern geopolitical imagination – irrespective of their geographical location. This is one of the insights emanating from so-called post-colonial studies, drawing attention to the worldwide penetration of dominant practices and understandings (such as those labeled by the term 'nationalism') and their naturalization into the routines of everyday life as 'common sense' and 'facts of life.' In a world of social practices rather than reified institutional forms, therefore, not only must states play by 'rules' established by dominant social groups that are active in all of them, the nature of rule-making presupposes that states (and other actors) are not *simply* coercive agents in a world of anarchy.

Two examples serve to illustrate why this argument matters. The first concerns the changing social and technological conditions for the military viability of states and their impact on the possibility of war serving reasons of state. The advent of nuclear weapons has meant that security now 'derives from the paralysis of states rather than from the exercise of state power, and from the acceptance of the impossibility of territorial violence monopolization rather than its pursuit' (Deudney 1995, 219). At the same time, the spread of easy-to-use conventional weapons (Kalashnikov machine guns and Stinger surface-to-air missiles, for example) has made it much easier for local populations to resist the designs of apparently more powerful adversaries. As Deudney puts it in summarizing the point about contemporary military practices: 'It is nearly impossible to protect territory from annihilation; but it is easier than ever to prevent conquest' (ibid.).

The second involves the contemporary decentering and deterritorialization (at a state level) of the means of production and communication. This reflects the opening up of the world economy for increased cross-border flows of trade and investment under American auspices during the Cold War. Recent developments in financial markets and information technologies, however, have accelerated changes in the ways people, places and states interact and how economic and political actors perceive these interactions. States (and others) must now *manage* these interactions. Though external coercion is a real possibility for the most powerful states, it is now of limited use when state policies must be concerned with attracting 'foreign' capital and gaining access to global flows of information. Contemporary economic practices, therefore, point towards the incipient creation of a *transnational–liberal* hegemony (see p. 59) in which territorial states are no longer the basic building blocks; they are being rapidly challenged by new spaces of networks and flows in which speed and access are more important than command over territory. Even in France, for example, often pointed to as a zone of changelessness by those skeptical of claims about an emerging world of flows challenging that of territories, recent neo-Gaullist and Socialist governments have supported a new European currency and a withdrawal of that state from many areas of life. It was General de Gaulle who said that France only exists because of the state, the army and the franc. His disciples and his political enemies now both stand watch over their dismantling.

STATEHOOD AND THE PROTECTION OF PROPERTY RIGHTS?

When states are situated in the context of a world of evolving social practices, they lose their exclusivity. But one historic role of states is thereby re-emphasized. This is that of the definition and regulation of property rights. The modern territorial state system has been associated from its origins in Europe in the seventeenth and eighteenth centuries with the framework for definitions of property rights (legal rights of ownership and use) without which global capitalism would not have been possible. States are never so 'sovereign', in the conventional sense of singular entities endowed with power monopolies within their territories, as when they are seen as definers and enforcers of property rights.

The third feature of state-centered accounts of the spatiality of power, therefore, is that they are silent as to the role that states have played in the growth of certain basic social practices of capitalism – defining and protecting private property rights – that have inexorably led beyond state boundaries in pursuit of wealth from the deployment of 'mobile property' (capital). The term *property* implies a fixity or permanence in place that modern territorial states have given a high priority to protecting. Consequently, much of the law in most Western states is devoted to establishing rights of ownership and access. But a home-territory also provides a base from which to launch attempts at acquiring property elsewhere. This requires that assets be reasonably liquid and transferable over space and across state boundaries. At a certain point, however, states endure a tension, what Ruggie (1993, 164) calls the problem of 'absolute individuation,' which can give rise to an 'unbundling' of territoriality when states effectively exchange control over economic flows emanating from their territories for increased access to flows coming from elsewhere. As a result, when increasing proportions of property are mobile beyond any one state's boundaries, individual states provide only a partial and tenuous protection for absolute property rights. Other geographical levels of governance and regulation become attractive, as was the case with the Bretton Woods system regulating international finance from 1944 to 1972 and is now (if signally less effectively) with the annual G-7 Summits between the leaders of the Big Seven industrial countries. But uncertainty as to future political actions and macroeconomic changes (tariffs, interest rates, etc.) also gives an incentive to property-holders to further spread assets around rather than leave them pooled up in one state.

This process is not new. Its origins go back to the merchant capitalism of the sixteenth century. What is new is the increased quantitative scale and the enlarged geographical scope of the mobile property now moving to and fro across the boundaries of the world's trading and investing states. In this context, states and firms have changed their orientation from free trade to what has been called 'market access' (Cowhey and Aronson 1993). The underpinnings of the world-trade regime that prevailed in the aftermath of the Second World War are being replaced by those of a regime in which a premium is placed on the openness of borders. 'Leakiness' in cross-border flows of goods and

Table 3.2 Pillars of the emerging 'market-access' world economy

Pillars of the free-trade regime	*Pillars of the market-access regime*	*Policy instruments*
Governance		
1 US model of industrial organization	Hybrid model of industrial organization	More reliance on bilateral and plurilateral negotiating forums
2 Separate systems of governance	Internationalization of domestic policies	Transparency, specialized rights of appeal, and self-binding behavior
3 Goods traded and services produced and consumed domestically	Globalization of services; eroding boundaries between goods and services	National treatment and tiered reciprocity for services
4 Universal rules are the norm	Sector-specific codes are common	Reforms of Voluntary Restraint Agreements, anti-dumping and subsidy codes
Rules		
5 Free movement of goods; investment conditional	Investment as integrated co-equal with trade	Rules of origin and new investment rules to ensure market-access, global antitrust
6 National comparative advantage	Regional and global advantage	Fair trade rules for procurement, standards, and R&D

Source: Cowhey and Aronson 1993, 237

investment and in firm multinationality has become a torrent of capital, trade and corporate alliances. Cowhey and Aronson (ibid., 237) contrast the nature of the old regime with that of the new one by identifying the six 'pillars' that they claim have underpinned each and the policy instruments associated with the new regime (Table 3.2). The policy instruments reflect an abandonment of classical state sovereignty in return for guaranteed rights of access to other states' territories. The world has moved away from the strict association of property rights and capital accumulation with state territoriality. A range of non-territorial factors now determine the competitiveness of firms in many industries: access to technology, marketing strategies, responsiveness to consumers, flexible management techniques. All of these are now the assets of firms, not territories. Firms grow through deploying their internal assets as successfully as possible. States now compete with one another to attract these mobile assets (property) to their territories.

Three aspects of the market-access regime are particularly notable with respect to the changing spatiality of power. One is the internationalization of a range of domestic policies to conform to global norms of performance. Thus, not only trade policy but also industrial, product liability and social welfare policies are subject to definition and oversight in terms of their impacts on market-access between countries. Another is the increased trade in services, once produced and consumed largely within state boundaries. Partly this reflects the fact that many manufactured goods now contain a large share of service inputs – from R&D to marketing and advertising. But it is also because the revolution in telecommunications means that many services, from banking to design to packaging, can now be provided to global markets. This represents a significant material challenge to the domestic vs. international distinction upon which the 'realism' of strictly territorial accounts of the spatiality of power relies. Finally, the spreading reach of transnational firms and the emergence of international corporate alliances have had profound influences on the nature of trade and investment flows, undermining the identity between territory and economy. Symptomatic of the integration of trade and investment are such frequently heard concerns as rules on international investment and unitary taxation, rules governing local-content and place-of-origin to assess where value was added in the commodity chains of globalized production, and rules involving unfair competition and anti-trust or monopoly trading practices.

TRANSNATIONAL LIBERALISM AND NEW SPATIALITIES OF POWER

Outlining the three main features of conventional accounts of the spatiality or geographical organization of power brings into focus their common deficiency: taking the coercive power of territorial states for granted as a fixed feature of the modern world rather than seeing it as the outcome of a number of historical contingencies. The contemporary 'unbundling' of state territoriality provides the most direct evidence for the reshaping of hegemony away from the state-centered practices of the previous epoch. This does not mean to say that territorial states are (finally) 'withering away,' only that they must now operate in a global context in which their interactions with one another must now take into account a changed military and economic environment. Indeed, in the absence of higher-level units for the enforcement of property rights and the delivery of public goods, states have a continuing and vital role to perform within the evolving world of networks and flows. For example, the deregulation of financial markets requires the deliberate action of governmental authorities. It does not simply 'happen.' During the Cold War between 1947 and 1990, the United States, in competing militarily and ideologically with the Soviet Union, sponsored an unprecedented opening of the world economy, partly to spread its political–economic 'message' and partly to take advantage of opportunities for its businesses. The net effect has been that markets have acquired powers heretofore vested in leading states. As this process has intensified and expanded, some localities and regions within states have been privileged within global networks of finance, manufacturing and cultural production to the disadvantage of others. The 'market-access regime' ties local areas directly into global markets. Successful ones are those which can enhance their position by increasing their attractiveness to multinational and global firms. A patchwork of places within a global node and network system therefore co-exists with but is slowly eroding the territorial spatiality with which we are all so familiar. Two consequences of this trend are illustrated in the rest of this chapter to give some empirical substance to the claim that a new geopolitics of power is in the offing: the explosion of non-territorial state political identities associated with global and local political movements (illustrated, respectively, by new post-national and local literary productions) and the increasingly decentralized world financial system, illustrated by the 'deterritorialization' of currencies.

FROM LITERATURE TO LITERATURES

One of the common assumptions of literary study is that of the historical conjunction between the creation of the novel as a literary form and the origins of the modern territorial state. Literary theorists from Lukács (1971) to Watt (1957) have claimed that the novel 'rises' (Watt's term) alongside the new state and the new classes it brings into existence, above all the new middle class or bourgeoisie, tying the idea of nation to that of state. Daniel Defoe's *Robinson Crusoe* (published in 1719) is sometimes credited with being the first novel, identifying the self-sufficient Englishman who is the hero of the story with a particular national space. His various traits also define the ideal–typical Englishman who is thereafter the subject of English novelistic discourse. The appeal to certain landscape images and historic forms of sociality (consider, for example, the novels of Jane Austen or Anthony Trollope) also served to confuse nation with state, the former involving the identity of a social group occupying a specific territory, the latter a patterned exercise of power within a bounded territory. The conflation (particularly strong in the English-speaking world) effectively underwrote the 'naturalization' of existing and prospective states as the right and proper representatives of the nations into which the world's population was seemingly divided (see Chapter 5).

The 'rise of the novel' also involved, however, the creation of a new category of literary production: Literature with a capital L. The new works demanded, in contra-distinction from older forms which were often oral in delivery,

> literacy, privatized, silent reading practices, and an elite to determine what constituted it. It depended on market capitalism for printed books, social practices that both permitted and limited literacy according to class and gender, and the invention and adjudication of taste by the individuals who constituted the elite.
>
> (Allen 1995, 99)

The study of Literature is still largely compartmentalized for reasons of linguistic specialization and competence into genres identified with specific state territories: Italian Literature, Irish Literature, etc. But two trends signal the extent to which the historic association of the novel with the modern territorial state is undergoing a significant stress. The first is the increased importance on best-seller lists around the world of so-called post-national novels. Examples of post-national novels would include Umberto Eco's *The Name of the Rose*, Salman Rushdie's *The Satanic Verses*, Milan Kundera's *The Unbearable Lightness of Being*, Gabriel Garcia Marquez's *One Hundred Years of Solitude*, and Margaret Atwood's *Lady Oracle*. These novels are sometimes deeply infused with particular national identities or readily associated with specific states but they all address themes that are concerned with placelessness (diaspora, existence loses its substance, etc.) and rapid movement across cultural worlds. Most importantly, at publication they appear almost simultaneously in many languages and are distributed on a global scale. Publishing is now one of the most globalized of activities. Place of publication has ceased to have much meaning for the publication of many works of 'fiction.' Some of the post-national novels have been made into films (another globalized industry) which reach an audience around the world that cannot or does not read novels at all. The appeal they have is that they usually show that the identities with which they are concerned are deeply problematic. In other words, they call into question the stability of the very identities older novels were intent on both representing and building.

They hold before people from elsewhere a mirror to their own identity in the exposure of someone else's national identity. For example, at the same time as he engages with the particular difficulties of his Czech protagonists at the time of the Prague Spring of 1968 and thereafter, Kundera's novel is about the worldwide victory of kitsch: of what cannot be thought or spoken of. As the literary scholar Beverly Allen (1995, 103) expresses the mirror argument, using a particularly evocative example:

> Consider the possibility, for example, that a non-print-literate person in Norway or the Philippines or the United States might see the film made from Kundera's novel. Each viewer's distance from the characters' negotiations of their own Czech national identity contains the possibility that the viewer's own national identity might take on a relative value, a sense of arbitrariness, even perhaps a tinge of virtual exchangeability in a world of floating rates of national identity exchange.

The worldwide circulation of the 'identity codes' contained within post-national novels, therefore, involves a shift away from the territorial nation–state as the total organizer of identity and towards a transnational space of engagement in which social identities are contingent and partial.

The second trend in literary production is the revival of local literature. In focus and content this literature pre-dates Literature and has existed in the shadow of national genres ever since the territorial nation–state took hold in Europe and the Americas. Its very existence has always weakened the monopolistic claims of national literary canons, particularly when expressed in a dialect or local language distinct from the national one. Characters are drawn from local rather than national 'types' and a world view is located in the thoughts and behaviour of people living localized lives outside of national or other 'larger' contexts. Local novels and poetry obviously appeal to audiences who are part of the worlds they portray. But they also, as with post-national novels from 'above,' call into question from 'below,' as it were, established social identities associated with particular nation–states when read by both insiders and outsiders. Allen (1995) uses the examples of contemporary Italian dialect poetry and the Glasgow novels of James Kelman and Alasdair Gray to show how local literature in its positive appraisal of regional, local and municipal identities exposes all identities as constructs rather than 'natural' categories. Particularities are what matter. In a passage in one of his short stories that deals explicitly with the issue of identity, Alasdair Gray (1993, 104–5) manages to convey very clearly what is at stake:

> No, my worst enemy could never accuse me of being a Scottish Nationalist. I don't approve of Scotland or Ireland – both Irelands – or England, Argentina, Pakistan, Bosnia et cetera. In my opinion nations, like religions and political institutions, have been rendered obsolete

> by modern technology. As Margaret Thatcher once so wisely said, 'There is no such thing as society,' and what is a nation but a great big example of our non-existent society?

So, what does he believe in?

> 'I am a Partick Thistle supporter,' . . . 'and I believe in Virtual Reality.' Do you know about Partick Thistle? It is a non-sectarian Glasgow football club. Rangers FC is overwhelmingly managed and supported by Protestant zealots, Celtic FC by Catholics, but the Partick Thistle supporters anthem goes like this:
>
> We hate Roman Catholics,
> We hate Protestants too,
> We hate Jews and Muslims,
> Partick Thistle we love you . . .

The very existence of local poetry, novels and short stories challenges the monopolistic claims of national genres of writing. When local novels enter into global circulation, as with the recent worldwide success of the Edinburgh novel *Trainspotting*, they feed into the rising tide of literature viewing terrestrial space, like the lives of that novel's protagonists, as profoundly shaken and fragmented. The flexing of local identities, based on historic claims to distinctiveness, works alongside the circulation of post-national novels, therefore, to call into question the fixity of the territorial spaces around which so much literary production has long been based.

MONEY AND STATES

The control and maintenance of a territorially uniform and exclusive currency are often regarded as one of the main attributes of state sovereignty. If a state cannot issue and control its own currency then it is not much of a state. Cohen (1977, 3) offers a concise statement of this view:

> the creation of money is widely acknowledged as one of the fundamental attributes of political sovereignty. Virtually every state issues its own currency; within national frontiers, no currency but the local currency is generally accepted to serve the three traditional functions of money – medium of exchange, unit of account, and store of value.

Currency has a further and vital symbolic role in underwriting statehood.

> As Keynes [the famous economist] understood, the creditworthiness of a nation's money is perhaps the primary evidence to the faithful (the citizens) that the ultimate object of their

> faith, the nation–state, is real, powerful and legitimate; it is the ultimate 'guarantor of value'.
>
> (Brantlinger 1996, 241)

Yet, over the past thirty years a number of trends have challenged the notion that every state must have its own 'territorial currency' (currencies that are homogeneous and exclusive within the boundaries of a given state). This does not necessarily portend a crisis for the state system as such so much as a challenge to that legitimacy of states which rests on claims to represent particular nations and associated national interests by means of control over singular currencies. Territorial currencies developed on a large scale only in the nineteenth century, once the 'Westphalian system' of states was already in place (Helleiner 1996). Symbolically, however, currencies (including the symbols found on coinage and bank notes) were important elements in establishing central state legitimacy long before the nineteenth century. Modern statehood was not achieved independently of processes of nation-building, even though 'state' and 'nation' can be distinguished analytically, the former referring to a set of institutions ruling over a discrete territory and the latter signifying a group of people who share a sense of common destiny and occupy a common space. The fact that the construction of territorial currencies was largely a nineteenth-century phenomenon, therefore, should not detract from the persisting linkage over many centuries between currency and statehood, however ineffective in practice that linkage often was.

Three monetary developments have begun to delink currencies from states in the way they were once largely mutually defining. The first is the growing use of foreign currencies for a range of transactions within national currency territories. The best known of these is the growth of the so-called eurodollar markets in London and other European financial centres. Others would include the development of off-shore financial centres such as the Bahamas and the Cayman Islands largely devoted to exchanging, sheltering and laundering foreign currencies. This trend is one part of that set of processes leading to global financial integration, at least among the world's richest economies.

The second is the emergence of projects such as that in the European Union to restrict or abolish national currencies in favour of supranational or world-regional currencies. In practice, the US dollar, the ECU (or European Currency Unit), the German mark and the Japanese yen have served as transnational currencies for some years. Much of world trade is denominated in one or other of these currencies, irrespective of its particular origins or destinations. Currencies such as the US dollar and the French franc have also come to dominate large regions beyond their borders, the dollar in Latin America and the French franc in parts of West Africa that formerly were part of the French Empire. Some of this is the result of internationally mandated economic reform, while some is more the result of local élites keeping their funds in 'harder' (more stable and reliable) currencies. The prospect of a 'Euro' or other European currency eliminating at least some of the existing European territorial currencies suggests that the process of transnationalization of currencies will intensify in years to come.

Third, and finally, a number of uses have appeared in recent years for 'local currencies,' forms of scrip and token money, that substitute for regular national currency. Such uses are often the result of experiments in local communities (e.g. Ithaca, NY and Montpelier, VT in the United States) and consumer cooperatives or tokens issued by firms for their products or services. As yet, they cannot really be seen as posing a major threat to territorial currencies. But they are often symptomatic of the lack of trust that territorial currencies now elicit in some quarters, perhaps as a result of the uses of monetary policies which have disadvantaged some groups (and localities) when currencies are rapidly revalued or high inflation persists which pushes people out of the official monetary economy and into a 'black' or underground economy where barter, trusted foreign currencies (such as the German mark in many parts of Eastern Europe) or local currencies prevail.

The deterritorialization of currencies, therefore, has three aspects to it: the explosion of foreign currency transactions within the territories of hitherto 'territorial currencies;' the rapid increase in the number of economic transactions involving supranational currencies; and the growing use of local currencies. None of these should be read as totally undermining existing territorial currencies. The continued erosion of territorial currencies will take place only if states continue to allow it. That there is still advantage in it for the most 'powerful' currencies means, however, that it will probably continue. As it deepens it could well gain a momentum that even the most powerful of states will find difficult to counteract.

CONCLUSION

At one time it made sense to some to see the path of history or social change as a series of 'stages' (as, for example, in Rostow's (1960) famous account of 'the stages of economic growth') inscribed upon state territories. Today, however, economic development and social change are increasingly determined by the relative ability of localities and regions to achieve access to global networks. In this context, understanding power as if it is attached singularly and permanently to state territories makes no sense. But the commitment to an unchanging spatiality of power retains considerable appeal. Not only does it allow for a restriction of politics to an unproblematic 'domestic' space, it also provides an attractive intellectual and political stability by equating space with the fixed territories of modern statehood which can then serve as a template for the investigation of other phenomena or as the basis for organizing political action. Putting state territoriality in question undermines the 'methodological nationalism' that has lain behind the workings of both mainstream and much radical social science. The major social sciences in the contemporary Western university – economics, sociology and political science – were all founded to provide intellectual services to modern states in, respectively, wealth creation, social control and state management. It is hardly surprising, therefore, that they find difficulty in moving beyond a world unproblematically divided up into discrete units of sovereign space. Political geography would seem to have less of an excuse. Putatively concerned with the spatiality of power, it has remained largely attached to a geopolitical imagination that has relied on seeing coercive power and the territorial state as twin markers of modernity. To correct this vision is a major challenge facing political geography at the close of the twentieth century.

RECOMMENDED READING

Biersteker, T. and Weber, C. (eds) 1996 *State Sovereignty as Social Construct*. Cambridge: Cambridge University Press.

Ruggie, J. G. 1993 Territoriality and beyond: problematizing modernity in international relations. *International Organization*, 47: 139–74.

Spruyt, H. 1994 *The Sovereign State and its Competitors: An Analysis of Systems Change*. Princeton, NJ: Princeton University Press.

Walker, R. B. J. 1993 *Inside/Outside: International Relations as Political Theory*. Cambridge: Cambridge University Press.

4

PURSUING PRIMACY

•

Within the modern geopolitical imagination there is a logical tension between the normative claim of an essential equality of 'statehood' between all states implicit in state sovereignty and international law, on the one hand, and the historical reality of dramatic inequality in power between them, on the other. This has been resolved intellectually and practically by seeing the normative claim of sovereign equality as an initial situation akin to a 'condition of nature,' with inequality and the resulting hierarchy of states as the result of an inevitable process of competition between states once 'social life' has begun. After reviewing the classic argument for why a hierarchy of states is a 'normal' condition in world politics, this chapter identifies the main axioms upon which this position relies, the historical–geographical conditions under which it made sense and the difficulties facing it under contemporary political–economic conditions.

THE SOCIAL ORIGINS OF GREAT POWERS

This position was foreshadowed in Hegel's famous discussion of lordship and bondage, in which nominally equal persons or 'selves' are seen as locked into an unequal and enduring hierarchy. This is so because a 'single consciousness' can know itself only through an other, even in a situation of radically unequal power. So, a lord is a lord only through a relationship of *mutual recognition* with a bond servant. The character of the relationship, however, is one in which 'the other' is sublated or annulled. Lordship derives from the conquest and negation of the servant. Recognition of the other is required, therefore, for its negation. A lord must have a servant who recognizes him as such to be a lord. In return, the lord must recognize the selfhood of the servant. Paradoxically, therefore, the mastery or *primacy* of one over another implicitly recognizes the separate selfhood of the other, even as the other is diminished. This is how primacy emerges within a world of nominally equal selves.

If states are substituted for selves, this is the understanding of what drives world politics that has predominated within the modern geopolitical imagination. The paradox between normative equality of sovereign states and the real inequality between them has been resolved by presuming an acquisition of primacy among states equivalent to the status allocation emanating from the lordship/bondage model. Higher-order states or Great Powers, therefore, emerge as a result of competition between states but can be accepted as such only if recognized by those states which are subordinated. Primacy depends in equal parts on successful competition and on subsequent recognition of that success by other states.

In the politically fragmented Europe in which the modern system of territorial states developed, military prestige was the main measure of competitive success. States formed a status system analogous to that found among social groups. Each sought to emulate the more prestigious and more modern states above them in the hierarchy. Hence, each state sought advancement by competing with other states. Modernization entailed the endless emulation of successful Great Powers by aspiring ones. Imperial expansion has been one strategy to claim or maintain the status of being a Great Power. Depending on how you look at, akin to whether you see a glass that is half-empty or half-full, every state's goal, therefore, is either to achieve primacy or to avoid subordination.

This view and the practices that relate to it depend for their veracity upon a historically fixed notion of statehood and a world in which modernity is in short supply relative to perceptions of backwardness (positions outlined in Chapters 3 and 2, respectively). Accepting the identity of states as self-sufficient actors with clearly bounded 'selves' is the most important move in both representing and practicing the pursuit of primacy. From this point of view, states have a 'first-person' identity akin to that of individual persons. This is why it often has made sense to talk of 'France' doing this or 'India' doing that, as if *each* were a capable actor in its own right. One of the most powerful metaphors in modern political theory is the idea of the state as an 'organism' or autonomous entity having a superordinate identity that cannot be reduced to any of its parts (its populations, social groups, etc.). This point of view, with its origins in ancient Greek philosophy, was given an explicitly biological cast in the late nineteenth century by authors such as one of the founders of political geography, Friedrich Ratzel, and reached its zenith with the Nazis. It opens the state to analogical reasoning; to treatment *as if* it were a person or a biological individual of some type. This organic reasoning became a vital part of the modern geopolitical imagination, particularly as it developed an affinity for 'naturalized' (fact-of-life) explanation in the late nineteenth century.

A second move sees 'social life,' once the initial state of nature has passed, as inherently competitive. Though the state provides a means for restricting conflict and encouraging cooperation within its territorial boundaries, beyond those boundaries is in essence a pre-social world in which the English political philosopher Thomas Hobbes's (1588–1679) 'struggle of all against all' goes on unabated. All economic and social welfare are seen as depending on expanding your state's capacity for violence relative to that of other states. This view has ancient European roots. The stress on the naturalness of danger and violence from strangers was used by the influential theologian St Augustine to justify war. He turned the Western Christian tradition away from pacifism by claiming that violence was intrinsic to human nature and could be channeled into righteous paths if it was used to convert the heathen and destroy the heretic. This political theology became widely accepted in Europe and underpinned the growth of 'just war' doctrines that were used to justify conflicts between Christian powers as well as with infidels of one kind or another. Although over the past two hundred years inter-state competition has had a political–economic cast, and in recovering the elements of the argument this chapter necessarily adopts an appropriately political–economic tone, the origins of justification for military and other competitiveness in the religious history of Western Europe suggest that it has a much more deep-seated cultural history that can only be touched on here.

The problem with the social argument for inter-state competition and the emergence of Great Powers is not only theoretical, as the next section attempts to show. It is also empirical. It no longer offers the same purchase on reality that it once seemed to. At present, a dynamic of globalization is prizing open even such previously closed or self-sufficient national economies as those of China, Russia and the United States, welfare states and government ownership are being undermined by privatization all over the world, and businesses increasingly operate in terms of world–regional and global markets (trends outlined in Chapter 3 in the discussion of states and power). In this historical context the imperatives of territorial and economic competition between states make less sense than they did when economic and political practices jointly encouraged them (as they did from the late nineteenth century until 1945; see Chapter 5).

The purpose of the rest of this chapter is to identify the historical–geographical conditions in which the pursuit of primacy did perhaps make some sense and why today this component of the geopolitical imagination increasingly does not. Before describing something of the historical geography of primacy, however, it is important to identify the major axioms upon which the presumption of a permanent or trans-historical pursuit of primacy rests.

AXIOMS FOR THE PURSUIT OF PRIMACY

The hierarchy of Great Powers can be seen as emerging out of a competition for primacy on the basis of two axioms concerning states and their attributes. One is that *relative* power differences between states cause states to compete with one another for relative shifts in power and status. The second is that competition between states takes place under conditions of international anarchy, i.e. conditions in which there is little or no return to cooperation, and winning is everything. These axioms have been adopted most formally and consistently by modern international relations theorists. But they also implicitly inform the modern geopolitical imagination as it has developed and changed down the years.

The focus on one's 'own' state and its security *vis-à-vis* the pre-emptive activities and potential depredations of others reflects the profound ontological insecurity (sense of loss of predictability and order) of people in the modern world. The geopolitical imagination has offered a reassuring response. Once security was no longer vested in a transcendental religious order with earthly enforcers, such as the medieval Christian Church, a substitute had to be found. Beyond the boundaries of the modern state, hypothesized such thinkers as the early modern Machiavelli and Hobbes, and the later Hegel, at least in typical readings of them (which is what matters here), lay an anarchic 'state of nature.' The geopolitical simplification of the world into 'friendly' and 'dangerous' spaces provided a practical means of giving order to this threatening and dangerous world. Sometimes a religious vocabulary has served to validate secular geopolitics in classical transcendental terms. Thus, one's most important opponent takes on a Satanic cast (America as the 'Great Satan' for Iran's Ayatollah Khomeini, the Soviet Union as the 'evil empire' to American President Ronald Reagan) or stands for the Anti-Christ (as in American-Christian fundamentalist understandings of the Soviet Union during the Cold War). This rhetoric, however, is usually a mask for the sense of a mysterious and alien threat coming from distant shores (and, after *Independence Day*, from other planets?). Anarchy 'out there' can only be countered by making sure that 'out there' does not come 'here.' Only primacy, a dominant global position underwriting national security, can guarantee that it stays 'out there.'

The first axiom for the pursuit of primacy refers to the idea that the power of individual states grows and declines at different rates, largely as a result of differences in rates of economic growth. In this construction, as some states gain power, others are losing it. Power is always gained or lost relative to others, rather than possessed or not possessed in an absolute sense. This presumes a 'global pool' of power of fixed volume that when drawn on by one state at an increasing rate diminishes that available to others. The international relations scholar Robert Gilpin (1981, 13) expresses this static storage-bin view of state power as follows: 'the differential growth of power of various states in the system causes a fundamental redistribution of power in the system.' The outcome, the historian Paul Kennedy (1987, xxii) asserts, is that, historically, 'relative economic shifts heralded the rise of new Great Powers which one day would have a decisive impact on the military/territorial order.'

Three behavioral assumptions must all be valid for relative shifts in economic growth to both make and shuffle the Great Powers. First, governments must see relative growth advantages as leading to improvements in the status of their states within the state system. They must orient themselves to achieving rates of growth that outnumber their competitors. Second, rather like stereotypical capitalists accumulating capital, states are never satisfied with small increments of status or even being first among equals. They always aspire to be 'Number One' or 'top dog.' Third, rising power is seen as invariably leading to increased military obligations and international commitments. As commitments expand beyond the capacity of a Great Power to finance them, primacy dissolves as adversaries with more modest commitments take their place at the head of the international league table. In sum, uneven rates of economic growth are seen as having cumulative effects that trigger the 'rise and fall of the Great Powers,' to quote Kennedy's (1987) phrase for the way in which primacy is first achieved and later undermined.

Much of the public discussion of Japan in the United States in the 1980s and early 1990s rests on these premises. The increased economic strength of Japan relative to that of the United States is seen as portending a shift in orientation from a trading state to a political–military one thus steadily replacing the United States as the world's Number One Power. That Japan's economy is still only half the absolute size of the American one excites little or no interest. Japan has been growing faster than the United States (though by the early 1990s this was not the case) and this trajectory supposedly bodes ill for the American position. American relative decline relates to the growth and presumed military expansion of Japan, therefore, rather than change in absolute capacities or change relative to rates of economic growth in the American past.

'Sporting' metaphors have been important in both pointing to the 'essential' differences between national potentials for world-class performance and in suggesting analogies that appeal to popular 'common sense,' a significant aspect of mobilization for war and other aspects of inter-state competition. Michael Shapiro (1989, 70) points out that comparing world politics to sporting contests serves the geopolitical purpose of emptying world space of any particular content: places lose their uniqueness and world politics becomes a type of strategic calculus, pure and simple. The world thus becomes like a gigantic playing field in which geographical location is equivalent to strategic location on a football field or basketball court. American presidents (particularly Presidents Nixon and Bush) have been particularly fond of sports metaphors as applied to world politics. 'Punting,' 'play selection' and 'end-run' were three football favorites that President Nixon often applied to foreign policy. They allowed a notoriously socially awkward man to appear as 'one of the boys,' engaging in dialogue with other sports-loving men. They also point to the obvious social–psychological linkages between sports as a type of sublimated war and the history of sports as a preparation for warfare. The Duke of Wellington may have started the trend in sports metaphors when he reportedly said that 'The Battle of Waterloo was won on the playing fields of Eton.' But, more critically, sports analogies allow a depoliticization of world politics. They encourage a sense of inter-state competition as a form of *pure* contest without moral or legal safeguards.

Zbigniew Brzezinski's book *Game Plan* is used by Shapiro as a convenient example of the genre. Supplying an 'integrated geostrategic framework for the conduct of the historical American-Soviet contest' (Brzezinski 1986, 8), the former national security advisor to President Carter implies in his use of a sports-contest discourse that

> 'what is simply involved is a struggle between two implacable forces that manifest a difference in ideology and historical motivation. He consistently represents the United States and the Soviet Union as 'two powers . . . fundamentally different.' And much of the book is addressed to aspects of this difference between the two contestants.
>
> (Shapiro 1989, 92)

The two contestants are depicted thus: one the 'real' man, the other the ersatz and lesser man. This obscures, of course, the costs imposed on others by the contest and the historical origins and interests served by continuing the contest into the indefinite future (Brzezinski's assumption at the time of writing). The image of contestants or competitors, therefore, naturalizes a process of global conflict between inevitable adversaries:

> Within the flattened narrative of the security contest, only national-level goals are implicated, and the 'threats' and 'catastrophes' that emerge as possible events touch only the interests of abstract entities situated in a geopolitical arena. The shape of the modern sports discourse and its use as a vehicle to figure world politics has the effect of situating us as spectators in a contest rather than as subjects in structures that create identities and locate us generally in a political economy distributing forms of danger.
>
> (ibid., 93)

The second axiom, that of international anarchy, derives from the view, implicit in the modern geopolitical imagination, that the state system is *inherently* conflictual at all times and in all places. In other words, if any one state gave up its ambition to be top dog another would be sure to replace it. This is why *relative* power is given such weight. If one state increases its relative power, then another state must either try to imitate its success or balance against it by allying with other states. The former is a so-called hegemonic strategy, the latter is a balance-of-power strategy. In either case a 'unipolar' situation is seen as always momentary. Sooner or later, states seeking upward mobility will act to reduce their subordination. The fact that Napoleon Bonaparte, Kaiser Wilhelm, Hitler, General Tojo, and Saddam Hussein seemingly acted on this basis to change the global *status quo* gives the perspective a set of persuasive historical examples for which prudence demands a response in kind.

The general perspective on international anarchy presumes, however, a universal condition of anarchy unaffected by space, time, or reflection. Anarchy is the outcome of *spontaneous self-serving action* by states. From this point of view, anarchy cannot be seen as the result of the projection of a set of particular values that underwrite particular relations of power and authority. It is more like a 'market' among states, in which each state is akin to an individual person pursuing its self-interest (whose definition is exogenous to the model) but devoid of any self-reflexivity about why it is doing so. It differs from the 'typical' market, however, in that spontaneity in this case does not give rise to a collective benefit and order but to inequality, hierarchy, and disorder.

This outcome results not only from the assumption of spontaneous action as states 'collide' with one another but also from the *lack of differentiation* among both types of polity (ways of organizing political life) and types of state system (ways of organizing states) that have existed historically. From this perspective, all states are presumed to be alike and part of the system of territorial states originating in Europe in the seventeenth and eighteenth centuries (the Westphalian System, named after the Treaty of Westphalia of 1648 which was based on the idea of single rulers for clearly bounded territories mutually recognizing one another's sovereignty). But there has been a whole variety of ways of organizing power geographically over historical time (many co-existing with one another during the years of the Westphalian system itself), from hierarchical but territorially mobile units such as nomadic empires (e.g. the Mongols) to non-hierarchical but territorially fixed ones such as sedentary tribes (e.g. Scottish clans) with the most modern (European) units combining hierarchical and territorial features, further differentiated one from the other by the relative importance of capital accumulation and coercion by dominant classes in their political organization (city–states, nation–states, and empires). There have also been distinctive types of state system in which different sorts of relationships have predominated between member units. The anarchic Westphalian system is the only type of system with presumed 'like-units' jockeying with one another for advantage. Other systems have been (1) suzerainties in which one unit has a superior status over others, e.g. US hegemony after the Second World War; (2) complex systems of hierarchical subordination and overlapping authority, e.g. the medieval European system in which the Pope shared power with a variety of monarchs and feudal rulers; the United States before the Civil War (what Deudney (1996) calls the 'Philadelphian system' in which the states had much more power than they did after the Civil War); and the European Union; and (3) classical systems such as that of Ancient Greece, involving heterogeneous but autonomous units. Such a variety of polities and state systems mandates against presuming a single type of polity or state system as historically inevitable.

The assumptions of spontaneity and of absence of functional differentiation among types of state and state system are equally crucial to the presumption of international anarchy. The 'inevitability' of hierarchy is veiled by the mask of spontaneously generated anarchy among like-units. This covers up ethical commitments that, as a consequence of the presumption of 'spontaneity,' need never be acknowledged. International anarchy's basis in 'spontaneity' serves to define a world of states that is the product of forces over which people have little or no conscious control. Inter-state competition just happens. Ironically, the purpose of the modern geopolitical imagination has always been to identify a rational basis for intervention in world politics even as world politics is construed deterministically. If intervention is possible, then there is also considerable scope for changing the values upon which intervention is premised. The assumption of inevitable anarchy is one of these.

The policy prescriptions to draw from the axiom of international anarchy have never been very clear, suggesting that its putative 'realism' is far from assured in all historical circumstances. Using the example of the contemporary United States after the Cold War, one reading of the situation would be that US governments should act to encourage multipolarity and a balance of power. American geographical isolation, the large national economy and possession of a nuclear first-strike capability guarantee the US a critical balancing role in a multipolar system. A different reading, however, would be that the US should act to reassert its primacy before it is too late to do so. From this viewpoint, multipolarity is seen as inherently unstable as other states attempt to achieve hegemony by replacing the United States at the top of the international hierarchy. In both cases the axiom of anarchy offers no guidance.

There is an interesting analogy between the claim for a struggle for primacy among states and the 'winner-take-all' character of the reward structure of American society. In winner-take-all markets tiny differences in performance translate into massive differences in rewards. The relentless stress on 'winners' and 'blockbusters' in athletics, show business, publishing, management, and universities (!), concentrates wealth and can encourage talented people into jobs that are less socially productive than other occupations they might pursue. Job markets are structured this way partly because of the multiplier effects of celebrity (record, advertising, and book deals, for example, follow highly rated relative athletic or acting performances), but also because of the obsession with being 'Number One' or coming first; an important measure of status in a society in which there are relatively few widely accepted indicators for social stratification. Rather like performers in a capitalist marketplace, states have been seen in the same light: engaged in a struggle for positional primacy that produces high returns for the successful few, but costly arms races and unproductive investments for all.

Although the obsession with coming first is such an obvious feature of American culture, it is by no means unique to it. The competitive element in the European state system has followed the extension of that system into the rest of the world, even if the mix of polities has been greater than the modern geopolitical imagination presumes. The rewards of success have appealed to political élites around the world, given that they are the ones who are most likely to benefit from enhanced international status.

The 'winner-take-all' analogy, however, has directly fueled the recent national competitiveness debate in the United States. This implies that the future course of competition will be economic more than military. Again, states, like individual persons or business firms in a market, are seen as in a struggle for survival in an anarchic world. Lower rates of aggregate economic growth relative to other Great Powers (such as Japan) are read as indicating an imminent loss of primacy just as in the past falling behind in the arms race was given a similar reading. This focus on relative gains and losses characterizes much of the recent controversy in the United States over the 'threat' from Japan. In a story of almost legendary proportions, Robert Reich (Secretary of Labor in the first Clinton Administration) had asked a number of groups of students, professional economists, investment bankers, senior State Department officials, and citizens of the Boston area which of the following scenarios for the US economy was most preferable: (1) one in which the US economy grows by 25 per cent over the next ten years, while that of Japan grows by 75 per cent; or (2) one in which the US economy grows by 10 per cent, while the Japanese economy grows by 10.3 per cent. Nearly everyone except the economists chose (2). Perhaps the magnitude of the difference drove some people away from (1). What is clear and disturbing, however, is that most people were willing to forego a much larger absolute increase in economic growth to prevent a larger relative advantage to Japan.

The idea of world politics as a horse race or sporting contest, in which finishing second is seen as finishing nowhere, has a persisting appeal. Indeed, US government economic policies are inspired by the same conception of relative gains. For example, US policies towards Japan on the export and manufacture under license of satellites and the FSX fighter aircraft in the 1980s were developed explicitly in terms of relative gains and losses. But, again, this is not simply some quirk of American culture. The obsession with relative gains from inter-state competition is integral to the modern geopolitical imagination. It has a worldwide constituency of political élites. So, as both a representation of 'how the world works' and as a practice among politicians, the pursuit of primacy has a continuing importance in the conduct of world politics, notwithstanding the shift from military to economic competition as the presumed underlying dynamic.

THE HISTORICAL GEOGRAPHY OF PRIMACY

Earlier chapters claim that the modern geopolitical imagination emerged in the historical context not only of Europe's overseas expansion but also of the progressive institutionalization of the European state system. In medieval Europe there were few fixed boundaries between political authorities. Communities were united only by feudal allegiance and personal obligation rather than by abstract conceptions of individual citizenship in a geographically circumscribed territory. As the identification of citizenship with residence in a specific territorial space became the central fact of political identity, sovereignty shifted from the person of the monarch to the territory of the state and its institutions. Concurrently, the emerging capitalist economy was given regulative boundaries by the pre-existing, if still weakly centralized, territorial states. Not surprisingly, territorial states came to be widely seen as the exclusive sources of 'activity' on the world map. Industrial capitalism reinforced this tendency because its dominant spatial division of labor was strongly organized on a state–territorial basis. In particular, the stock Hobbesian view of individual persons locked in a struggle for survival in a hostile environment was easily extended to states. In this construction, the territorial state could be seen, as it is in the political economy of mercantilism, as a single abstract individual located in an environment of global anarchy.

The connection between this increasingly state-centric world and the modern geopolitical imagination lies in the sharp division that is drawn between the 'civic order' prevailing within state boundaries and the struggle for power between states beyond them. Just as major political ideologies provide the competing frameworks for political competition within states, the geopolitical imagination has provided the frame of reference for organizing state activities beyond them. Places near and far can be classified and related to the hierarchy of states: where are the main competitors? Where are the best possibilities for enhancing competitive position? Where are the main challenges to the established hierarchy? Maps of threats and dangers can be drawn based upon the inter-state pecking order at any particular time, identifying both its weakest and its strongest links.

Not all states have had the same capacity to realize their geopolitical visions. The so-called Great Powers of every era have been able to inscribe their particular geopolitical imaginations onto the world as a whole. State power has involved the capacity for undertaking action by states in the conditions of a particular era. This presupposes conventions of meaning about state behavior that are shared by all parties and arise out of the actions of states and other actors. In this construction, 'hegemony' refers to the norms and rules governing world politics accepted by dominant social groups and classes. The meanings are those diffused by the most powerful states, sometimes they emanate from single 'hegemonic' states; indicating the material reality of primacy in certain historical settings, such as that of Britain in the period 1815–75 and that of the United States after the Second World War (see Chapter 5). But there have been periods, such as that from 1875 to 1945, in which the dominant representations and practices of states have not had a single origin or sponsor but have been more widely based among political élites in a range of states. Changing economic and technological conditions are of vital importance in determining the ebb and flow of candidacies for primacy (either in the sense of territorial or economic dominance over other states or as the primary 'rule-writer' in world politics). Particularly significant has been the relative ability of different states over time to capture the positive externalities from new technologies with novel economic and military applications.

One scheme (see Chapter 5 for more details) identifies three periods from 1815 until 1990 in which distinctive combinations of economic, technological and political factors produced very different conditions for primacy. The first (1815–1875) rested on (1) a balance of power among a set of national states and territorial economies in Europe (plus the growing United States) with (2) an expanding tendency towards investment and trade organized non-territorially outside of Europe with British institutions as the main organizing agents. The second period (1875–1945) was one in which a number of imperial rivalries produced a struggle for primacy by means of territorial control over chunks of the world economy. It is in this period that the transcendental 'urge' of Great Powers to subjugate one another could be said to have been practiced in its most open form. Initially, the competition between imperial states (such as Britain and France) led to the creation of specialized colonial economies. These were organized largely around exclusive zones tied to particular Great Powers. This gave rise to enmity from those states (such as Japan and Italy) closed out of the enterprise. Failures in economic regulation at state and global levels (manifested in the Depression of the 1930s) encouraged a further retreat into territorial blocs. The final outcome was a struggle for supremacy between competing ideologies over how best to organize and, on one side, move beyond the blocs into an alternative type of political–economic organization.

The third period (1945–1990) saw two states emerge as the main contenders for global primacy: the United States and the Soviet Union. The American agenda was for the United States to perform as an 'international' state preventing the re-emergence of the blocs that many influential Americans (and others) believed produced or exacerbated both the Depression of the 1930s and the Second World War. Three strategies were pursued to further this agenda. The first was the demilitarization and economic reorientation of Japan and Germany. The second was the containment of the Soviet Union and its state-socialist model of economic development through military alliances (such as NATO) and political–military intervention (as in Korea and Vietnam). The third was the creation of a set of institutions to project American practices and ideas about political–economic organization at a global level. These included the United Nations, General Agreement on Tariffs and Trade, the International Monetary Fund, the World Bank, and such international agreements as the Bretton Woods Agreement of 1944 governing international monetary relations. By the 1960s these strategies had produced the beginnings of a globalized world economy in which many states became progressively internationalized. The costs of global military confrontation and the failure of the Soviet economy to keep up in information and communication technologies, worked with other factors (such as social movements) to undermine the Cold War order. *Pax*

Americana, designed to promote economic interaction between non-socialist national economies, encouraged a degree of globalization that effectively dented the powers of almost all territorial states, including the United States itself and its main adversary, the Soviet Union.

The 'pursuit of primacy,' therefore, is not simply an ideological construct that has been 'exposed' as such by the recent destabilization of political boundaries at the end of the Cold War period. It does seem to have provided a socio-economic purpose to statehood under certain conditions. But it is difficult to accept it as a timeless attribute of world politics. In the first place, the pursuit of primacy has taken very different forms in different historical periods. The mix of military and economic elements, for example, has varied significantly over time, partly as a result of the shifting possibilities for mercantilist and open-trading economies. More importantly, the way the most recent period has ended with the erosion of many state powers may signal the beginning of its demise. What now seems increasingly clear is that as a feature of the modern geopolitical imagination, the transcendental conception of the pursuit of primacy involves the projection of a historically specific set of practices and representations onto world politics in general. With the value of hindsight, the two geopolitical assumptions that underwrote the plausibility of the pursuit of primacy as a fundamental element of the modern geopolitical imagination appear now as historically specific rather than generally valid: (1) that power flows from advantages of geographical location, size of population and natural resources combined in a territorial mode of production; and (2) that power is entirely an attribute of territorial states that attempt to monopolize it in competition with other states.

In light of recent trends there are five specific ways in which the pursuit of primacy fails to pass a 'test' for trans-historical significance. First, a focus on relative power only makes sense in a world in which there is a constant threat of a war of attrition between Great Powers. In such circumstances constant preparation is required to achieve or maintain a relative advantage. The possession of and ability to deliver second-strike nuclear weapons render this vigilance obsolete. What matters in the nuclear era is not relative advantage, it matters little (how many times over can you destroy them?) but absolute ability to inflict deadly damage (see Chapter 3). A single nuclear device planted in a major city by a 'weak' state (or stateless terrorists) could provide such an advantage without the immense cost associated with developing new weapons and their costly delivery vehicles. Indeed, in this situation a focus on relative advantage is dangerously misleading: it could give rise to a self-fulfilling prophecy as preparing for nuclear war makes such war more rather than less likely. The advent of nuclear weapons, therefore, irrespective of the political character of governing regimes (democratic/authoritarian, etc.), reverses the logic of war-preparedness first made in the modern era by Machiavelli. More need no longer be better. Nuclear weapons appear to make governments cautious, whether or not they are the United States or the Soviet Union (stateless terrorists are another thing entirely). The United States is in the situation (1997) of accounting for 40 per cent of world military spending, seven or eight times that of Russia, France, Germany and Britain, the other military Great Powers. By one calculus this is absolute primacy. But even those proposing continued high levels of military spending, such as the US Secretary of Defense, cannot find military threats that justify them. Only North Korea, Iran, Iraq, and Syria are singled out as potential adversaries. Other Great Powers, presumably those most likely to be potential primates, do not figure at all in threat assessments. Other Great Powers are finding other things to do with their money.

Other features of contemporary military organization and technology are also making relative power less significant. One is the increased role of partisan or terrorist fighters in world conflicts. These are often stateless warriors, battling for this or that cause but without the infrastructure or the typical goals of a state-based military machine. Relatively simple but nevertheless devastating technologies (bombs, Stinger missiles, etc.) provide small mobilized groups with the means to neutralize or confound much more powerful states (consider the IRA in Britain or the ETA in Spain, for example). Another is the access to the information networks that are at the heart of military preparedness. Not only can small states and non-state groups potentially gain entry to and destroy command-and-control systems (if they have talented hackers and a good computer), they can also engage in 'netwar' (as have the Zapatista rebels in southern Mexico): using the Internet to send messages to sympathizers and non-governmental organizations so as to publicize and undermine state counter-insurgency operations. This is information warfare with a new twist, one that benefits small marginalized groups more than the big battalions of the Great Powers.

Second, economic growth and prosperity among the Great Powers have been enhanced since 1945 by increasing international trade and investment. The focus on relative gains and losses reflects a mercantilist economic ideology in which national economies are seen as equivalent to containers that can be filled only at one another's expense or through territorial expansion. Today, rates of economic growth are much higher when economies are open rather than closed. This is one of the lessons of the great Soviet collapse of 1990–92. Access to global markets is now the principal pre-condition for sustained economic growth. In turn, increased economic interdependence increases the incentive for resolving disputes non-militarily. The assets of businesses and individuals from particular states are no longer tied exclusively to the state's territory. Such foreign ties and stakes are a force for symmetrical interdependence when they are shared by all parties. This reduces the element of conflictuality and increases the return to cooperation.

One obvious criticism of the competitive primacy model, of course, is that states are not at all like competing firms in an economic model of 'perfect competition.' For one thing, as noted previously, they are not all alike, whereas competing firms often are. Neither is it clear that most states actually have single goals analogous to increased market share or greater profits. They (like persons and firms) can also benefit from cooperation with one another. Military alliances and regional trading arrangements are examples of ways in which such benefits have long been realized. Another criticism, made by advocates of the advantages to all parties involved of free trade and comparative advantage, is that national economies often suffer when restrictions are placed on imports and exports. In this view, although special interest groups concerned about the prospects of this or that industry may like to persuade everyone otherwise, entire national economies are *never* in substantive competition with one another *when there is relatively open trade* (for an argument along these lines see the economist Paul Krugman's (1995) reply to the historian Paul Kennedy, 1995). There are returns to specialization that can only be captured through either firm organization (e.g. vertical integration) or local/regional concentration. As an important trend in the world economy is away from the former to the latter in leading economic sectors (e.g. informational technologies, aerospace, producer services), national economies (as opposed to more localized ones) are even less likely to be in 'competition' than they were in the past when industries were more organized on a national rather than a transnational basis. An ironic confirmation of this trend can be found in comparing the title of the management guru Michael Porter's most well-known book, *The Competitive Advantage of Nations* (1990), with its contents. The book is in fact about the increasingly localized and regionalized basis to economic performance within industrialized states; about what Allen Scott (1996) calls the new 'regional motors' of the world economy.

Symptomatic of the internationalization of states is the growing concern of both states and firms with 'market access' (as described in Chapter 3). Though states continue to control the rules and structures of their own national economies, through formal international agreements and informal but explicit bargains they also grant access to foreign competitors. Classic free trade and investment policies, therefore, are being extended into the areas of industrial policy and domestic regulation to facilitate cross-border business collaboration. This is an important consequence of the recent rounds of the GATT and the growth of regional trading blocs such as the European Union and NAFTA.

Two measures of this growing internationalization are the increasing difficulty of accounting for trade in terms of transactions between national economies and the growing number of products without clear 'national origins.' The tremendous expansion of foreign direct investment and sub-contracting means that the national trade balances for countries such as the United States and Japan contain within them large numbers of items manufactured overseas by firms operating from within their confines. For example, according to the official books, the American trade deficit in 1986 was $144

Table 4.1 Conventional and ownership-based trade balances, United States (1986) and Japan (1983), in US$ billions

		United States	*Japan*
Exports	[1]	224	146
Less intra-firm transfers		123	60
Plus local sales to foreign MNCs		267	3
Plus sales by home-owned MNCs abroad		777	150
Equals total foreign sales	[2]	1,145	239
Imports	[3]	368	114
Less intra-firm transfers		191	65
Plus local purchases from foreign MNCs		445	58
Plus purchases by home-owned MNCs abroad		446	90
Equals total foreign purchases	[4]	1,068	197
Conventional trade balance	[1] – [3] =	–144	+32
Ownership-based trade balance	[2] – [4] =	+77	+42

Source: Julius 1990, 81

billion. But if you include the activities of US-owned firms abroad and foreign-owned firms in the US, the huge deficit disappears and becomes a surplus of $77 billion. The two figures for Japan are closer, reflecting the lower rate of foreign investment by Japanese industrial companies (Table 4.1).

What is now traded is also less often final products than components, materials and specialized research, design, fabrication, advertising, and financial services. One example comes from that quintessential American company General Motors, whose well-being was once considered to parallel exactly that of the US as a whole:

> When an American buys a Pontiac Le Mans from General Motors, for example, he engages unwittingly in an international transaction. Of the $10,000 paid to GM, about $3,000 goes to South Korea for routine labor and assembly operations, $1,850 to Japan for advanced components (engines, transaxles, and electronics), $700 to the former West Germany for styling and design engineering, $400 to Taiwan, Singapore, and Japan for small components, $250 to Britain for advertising and design services, and about $50 to Ireland and Barbados for data processing. The rest – less than $3,000 – goes to strategists in Detroit, lawyers and bankers in New York, lobbyists in Washington, insurance and health care workers all over the country, and to General Motors shareholders all over the world.
>
> (Reich 1991, A6)

A third failure of the pursuit-of-primacy thesis is that American 'hegemony' has been qualitatively different from that of previous eras. It has been institutionalized globally through a large number of agencies and has had a profound cultural influence. The demise of the Soviet Union has left the United States as the only Great Power with a global message: mass consumption, personal liberty, private property, markets, and electoral democracy, to name just the most obvious elements of this message. Of particular significance, US efforts at enmeshing other states into international 'regimes' of one kind or another covering a wide range of substantive issues from trade to fishing rights and environmental degradation have had the effect of emphasizing collective more than unilateral approaches to resolving conflicts.

Fourth, the number of truly Great Powers is now very small and the incentives for cooperation between them are very high. Under conditions of interdependence, each state has a stake in seeing that the others adhere to common rules governing trade, security, and investment. A diffusion of power or pluralism, more than a multipolarity among singularly 'selfish' states, is now the emerging trend among the Great Powers. This points up the social nature of international relations. Conventionally, and reflecting the *idée fixe* of the pursuit of primacy, relations between states are seen as inherently conflictual, as in the Prisoner's Dilemma game (Figure 4.1). In this game, both actors have identical preferences, both acquisitive and precautionary. Each must act without information of how the other will act. Each can benefit without making the other worse off by cooperating (+20 apiece). But the greatest return comes from defecting before the other does (+40). If both opt for defection both experience losses (–20 apiece). Without learning and confidence-building measures this is the presumed likely outcome. In this game a two-party conflict will not produce cooperation since each party believes it can always get more by not cooperating (see the higher potential returns to non-cooperation than to cooperation; potential losses are, of course, commensurately higher). This presumes, however, that each state is a self-sufficient actor without divided interests, there is a set of restrictive informational constraints and there is a lack of institutional and communication channels for accomodating differences of interest. Recent experience of international negotiations calls each and every one of these into question.

Fifth, and finally, there is a growing élite and popular consciousness of global interdependence. Slowly and fitfully a sense of a common global fate has started to grow. This is encouraged by the spread of access to information about previously unknown and distant places. But it is more directly the result of a growing sense of linkages and dependencies in a more interdependent world. This need not translate into an immediate bonding with neighbors, let alone distant others. It does indicate, however, the arrival of a consciousness of the impact of distant events (oil spills, burning forests, refugees from communal strife, nuclear winter, etc.) on the prospects for life at 'home' which do

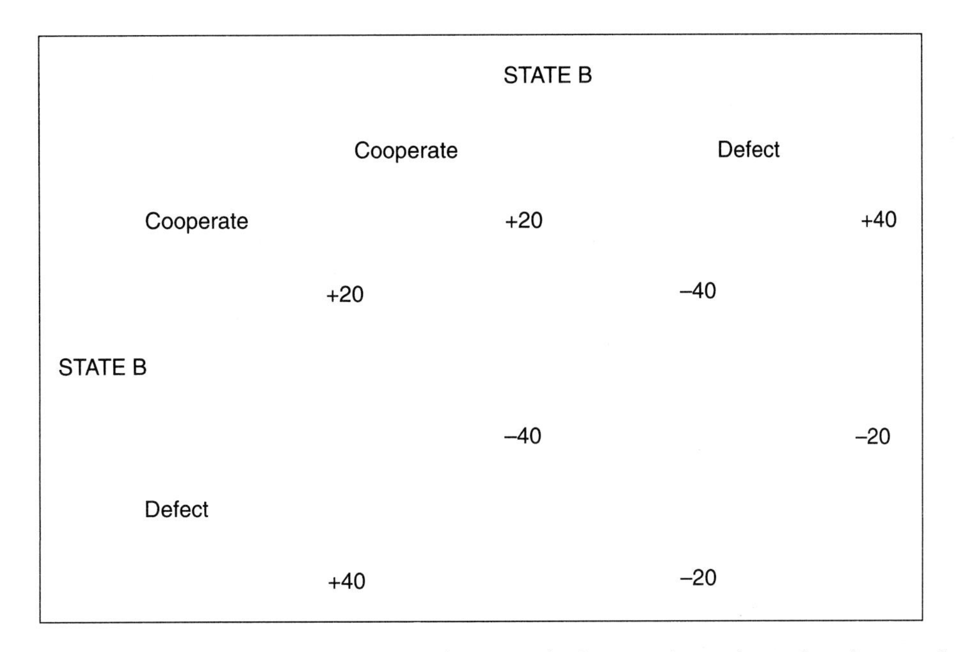

Figure 4.1 The Prisoner's Dilemma game. In this game, both actors have identical preferences, both acquisitive and precautionary. Each must act without information of how the other will act. Each can benefit without making the other worse off by cooperating (+20 apiece). But the greatest return comes from defecting before the other does (+40). If both opt for defection, both experience losses (–20 apiece). Without learning and confidence-building measures this is the presumed likely outcome.

not have simple military solutions to them. The success of transnational activists in generating sanctions against apartheid South Africa in the 1980s is a good example of how both a non-state-based movement and a concern for norms of conduct by states rather than national interests are now an important part of world politics. The recent upsurge in ethnic conflicts might suggest exactly the opposite. Indeed, there obviously are situations where group identities and interests are so polarized and threatened that peaceful resolution of conflict is next to impossible. But new spatial practices involving increased trade and population movement can be expected to slowly force a sense of common fate even as they also generate enmity and conflict. Some parties to the Israel–Palestine dispute, undoubtedly one of the world's most intractable conflicts, increasingly see a future in which each side will have to depend on resources and skills provided by the other. Why? Because what neither side can do 'is conjure up a pair of coherent nation–states' (Avishai 1995). Each population is now too intermingled with the other.

BEYOND THE PURSUIT OF PRIMACY

In the contemporary world, therefore, the pursuit of primacy has limited possibilities. Of course, this world might not last. Elements of opposition to the disruptive consequences of globalization are increasingly visible. The emerging transnational liberal order can be seen as detracting from rather than augmenting the material conditions of life for both the 'average' and the poorest people in countries where this process has gone the furthest. The 'hollowing out' of the US economy, due to technological change and the globalization of production more than to the direct effects of trade, has produced a deepening inequality of incomes between both social strata and regions. This could lead to increased success on behalf of calls for a reterritorialization of the US economy through protectionist measures (such as tariffs, quotas, content requirements, and product liability regulations). The calls are already there; from those concerned about immigration to those worried about the lack of jobs for the unskilled and the uneducated.

Unfortunately, in the world economy as it is now organized this would probably only lead to greater disinvestment from the territorial economy of the United States. This is what happened in France in 1983 when the Mitterrand government nationalized assets and put restrictions on banks and was immediately faced by a massive outflow of capital. What is more likely to meet with success is if societies collaborate multilaterally to regulate the socially disruptive consequences of globalization. Just as in the past social democratic movements arose to counter at the state level the socially damaging impacts of unfettered markets, so in the future 'multilateralism will become an arena of conflict between the endeavour to buttress the freedom of movement of powerful homogenizing economic forces, and efforts to build a new structure of regulation protecting diversity and the less powerful' (Cox 1992, 177). Whatever happens, therefore, a return to the pursuit of primacy seems an unlikely recipe for anyone's success, except those interests in the militaries and defense industries whose careers and profits have come to depend on it.

The world in which states must now operate has changed beyond recognition within one generation. A variety of economic and political processes have encouraged this. Prime among them have been the globalization of production and finance, the informational revolution in computing and telecommunications, the internationalization of states enmeshed in regimes of all kinds, massive migrations from poorer to richer countries, and the spread of a 'world culture' emanating from the Hollywood studios and CNN.

The most vital geopolitical consequence of the displacing of state boundaries by the flows of people, goods, and capital moving between local nodes in global networks has been the undermining of conventional processes and understandings of spatial *hierarchy*. The pursuit of primacy presupposes a world map in which state territories can contain within their boundaries the lion's share of the most profitable transactions that they generate. The returns from these transactions can then be invested in political–military adventures of one kind or another designed to establish the state's primacy or high ranking among 'comparable' states. For much of the last two hundred years this has largely been the case. In this respect the modern geopolitical imagination's commitment to the pursuit of primacy reflected the dominant practices of the Great Powers whose world it sought to understand and expand. American hegemony after the Second World War, however, undermined the pursuit of primacy in building the foundations of a world economy that has now escaped from the territorial confines of existing states.

The pursuit of primacy requires a hierarchy of *settled political identities* like those implicit in Hegel's model of lordship and bondage and in conventional understandings of the origin of the hierarchy of states. Changed historical–geographical conditions draw attention to this requirement. The trend from a geography of boundaries to one of flows that is integral to globalization, therefore, signals a crisis of representation for the modern geopolitical imagination. There is a crisis in the intellectual authority of established conventions about the behavior of the Great Powers: the main active figures in classical geopolitics. Not the least of these is that of the pursuit of primacy. But how did the modern geopolitical imagination work in practice between Hegel's day and ours? This is the subject of Chapter 5.

RECOMMENDED READING

Collins, R. 1986 *Weberian Sociological Theory*. Cambridge: Cambridge University Press.

Ferguson, Y. H. and Mansbach, R. W. 1996 *Polities: Authority, Identities, and Change*. Columbia, SC: University of South Carolina Press.

Kennedy, P. 1987 *The Rise and Fall of the Great Powers: Economic Change and Military Conflict from 1500 to 2000*. New York: Random House.

Krugman, P. 1996 *Pop Internationalism*. Cambridge, MA: MIT Press.

McNeill, W. H. 1982 *The Pursuit of Power*. Chicago: University of Chicago Press.

Porter, M. 1990 *The Competitive Advantage of Nations*. London: Macmillan.

Reich, R. 1991 *The Work of Nations: Preparing Ourselves for 21st Century Capitalism*. New York: Knopf.

Shapiro, M. J. 1989 Representing world politics: the sport/war intertext, in J. Der Derian and M. J. Shapiro (eds) *International/Intertextual Relations*. Lexington, MA: Lexington Books.

5

THE THREE AGES OF GEOPOLITICS

•

The purpose of this chapter is to show how three specific epochs or eras of geopolitics developed over the period from the early nineteenth century to the 1980s, built upon the interaction of changing material conditions with the 'principles' of the modern geopolitical imagination described in previous chapters. Even though the geopolitical imagination of each 'age' is distinctive in many respects, there are continuities as old themes are recycled in new contexts. The three discourses or modes of representation I discuss are referred to as civilizational geopolitics, naturalized geopolitics, and ideological geopolitics, respectively. World politics in each of the epochs has been organized around the characterizations of space, places, and peoples defined by these modes of representation.

CONTINUITY AND PERIODIZATION IN THE MODERN GEOPOLITICAL IMAGINATION

There is an obvious continuity running through the modern geopolitical imagination in a number of ways. One is the continuing use of a language of spatial difference expressed in terms of a temporal metaphor (modern/backward). Another is the world visualized as a whole as a field of reference for inter-state relations. A third is the exclusive role of territorial states as the actors in world politics. Finally, there is the pursuit of primacy by Great Powers as the motivating force behind world politics. However, the idioms and contexts of usage have changed dramatically over time. There is considerable intellectual danger in assimilating all geopolitical discourse into an overarching continuity flowing from the Renaissance to the late twentieth century and arbitrarily selecting themes or linguistic forms without attention to the contexts (historical and geographical) in which they have arisen. After all, the technological, economic, and social conditions which constrain and enable both representations and practices have changed significantly over the years. For example, the world economy today is organized completely differently from how it was during the heyday of European colonialism in the late nineteenth century.

A periodization of geopolitical discourse obviously simplifies a more complex flow of representations and practices. Each 'period' has within it the seeds both of its own demise and of subsequent periods. The dating of periods is tentative and contestable. Each could be subdivided to represent the inevitable shifts and turns in understandings and practices. The issue periodization addresses is that of dominant trends, understandings, and associated action and how these differ over time. Such a historicist conception does not deny the continuities identified in previous chapters. Rather, in each period the various 'principles' intersect in different ways to produce a different 'mix' of geopolitical representations and practices. This gives rise to different hegemonies, dominant sets of rules governing world politics, in different epochs (see Introduction). These correlate with the economic, technological, and social trends noted in Chapter 4 as giving rise to the threefold periodization: first, the early nineteenth century (1815–75); second, 1875–1945; and third, 1945–90. But, it should go without saying, given the arguments and claims of earlier chapters, that the dominant geopolitical imagination flowing through these periods has older roots in the growth of state-formation and capitalism in early-modern Europe and European expansion into the rest of the world. Neither are the hegemonies intrinsic to the different eras free of contradiction or contestation. Indeed, it is out of their internal contradictions that old hegemonies are undermined and new ones arise.

CIVILIZATIONAL GEOPOLITICS

In the late eighteenth century a *civilizational geopolitics* emerged as part of the reaction to the 'struggle for stability' in a Western Europe that had lost its cosmic center based on a common Christianity at the time of the Wars of Religion. This was an element in what Toulmin (1990, 170) has termed the attempt at constructing 'a more rational Cosmopolis, to replace the one lost around 1600.' This version of the modern geopolitical imagination peaked in influence in the first half of the nineteenth century. Before that time its various principles were not engaged with one another in a common whole in the way they were to be after 1815.

The international political economy from 1815 until 1875 was characterized by a European Concert in which no one state 'laid down the law' for the others within Europe and an emerging British economic dominance in much of the rest of the world. The Concert was designed to limit the revolutionary impulses emanating from the French Revolution of 1789 and its aftermath. Wars to foment revolution or to acquire territory were seen as threats to the 'balance of power' between the dominant states or Great Powers. At the same time, Britain's head start in economic growth as a result of its Industrial Revolution produced a different interaction with the other continents than that of the other European states (save for the Netherlands). The search for markets for manufactured goods and the availability of capital for overseas investment produced pressures for external expansion (both colonial and more generally) that set Britain apart. As the rest of Europe became caught up in national economy-making, a process largely completed in Britain, British businesses and political leaders created a worldwide network of trade and financial flows.

At this stage in the growth of world capitalism, British centrality was widely seen by European political élites as serving not only a national interest but a 'global' interest as well. The British national economy became the locomotive of the world economy. The recycling of capital from worldwide investments made London the world's main financial center. The main political interest of financial capital lay in preventing economic disruption in places where it had investments. Not surprisingly, British governments also came to value 'stability' and often intervened militarily outside of Europe when the *status quo* was under threat. With the economic success of the German and American economies, the British economy, still largely committed to older technologies, was faced with a dilemma. On the one hand, it was faced with increased competition in manufactured goods and, on the other, it had a relative advantage over the others in its access to the rest of the world. Consequently, after 1875 Britain turned away from Europe and the United States (where British capital had major investments) towards its empire and those world regions where its hegemony was apparently more secure. This gave impetus to the final collapse of the Concert of Europe, already threatened by the emergence of new states in Europe (such as Germany and Italy) without any commitment to it, and the emergence of a global system of competing empires as others, from Germany, France, the United States and Japan, attempted to follow in Britain's footsteps. But it was within Europe that the most important challenge of the age arose. The political order of civilizational geopolitics was that defined by aristocratic and bourgeois élites. In its main features it was designed to tame and repress the nationalism generated by the French Revolution. But as the nineteenth century wore on, a new conception of statehood based on the creation of *nation–states* came to the fore. These were states built upon cultural divisions and particularities. In the identity of state with nation, territorial sovereignty was fused with the fate of the nation. The 'interests' of peoples were rigidly territorialized as the number of states in Europe tripled. This new system of states had no place for civilizational geopolitics.

It was in the political–economic context of 1815–75 that civilizational geopolitics was most fully established, even though the 'principles' it drew on had older roots, as described in previous chapters. Its main elements were a commitment to European uniqueness as a civilization, a belief that the roots of European distinctiveness were found in its past, a sense that though other cultures might have noble pasts with high achievements they had been eclipsed by Europe, and an increasing identification with a particular nation–state as representing the most perfected version of the European difference.

The idea that the earth's land area is divided into separate 'continents' was first proposed by ancient Greek geographers who identified three continents – Europe, Asia, and Africa – bounded by the Mediterranean Sea and the Rivers Nile and Don (see Chapter 1). Although later geographical knowledge suggested that Europe and Asia were not clearly demarcated from one another by a significant body of water, the division of the world into continents persisted because the concept of Europe itself changed. From a physical–geographical region the geographical realm of Europe was transformed into a cultural region. This happened as the Christian Church abandoned its claims to universality and defined a much more narrowly circumscribed Christendom. The Arab and, later, Ottoman 'perils' gave particular credibility to the sense on the part of vulnerable Europeans of a profound chasm between the familiar world of Christian Europe and the exotic world of the Moslem Other (see Chapters 1 and 2).

Imaginative maps showed Europe as a 'queen' among the continents, complete with orb and sceptre (Figure 5.1). This illustrates not only a sense of difference but also an emerging sense of superiority. This was reinforced by the European voyages of 'discovery' which demonstrated the self-evident initiative, vision, and zeal of Europeans. Over the next centuries, the feeling of superiority 'gradually hardened into an inflexible conceit that held Europe to be the most civilized and best governed of all the world regions' (Bassin 1991, 3).

The basis for this conceit was not simply comparison with benighted non-Europeans but, more importantly, a sense of a distinctive past (described in Chapters 1 and 2). The rest of the world was 'available' for use by Europeans because their history destined them for Greatness. It is no mere coincidence that the early nineteenth century saw an obsession among European élites with the 'examples' of ancient Greece and Rome. This was particularly marked in the United States; in the minds of its leaders very much an extension of European ideals beyond the physical boundaries of Europe. The 'founding fathers' of the United States looked to Rome for their model; they thought of Athens as ruled by mobs. In the early nineteenth century this was displaced by a much more popular Greek Revival which affected everything from political oratory to town names and house design. Throughout Europe a 'classical' education became a prerequisite for social and political success. Contemporary events were understood in terms of classical referents.

The dichotomy drawn between Europe and the other continents was reinforced by the frequent combination of an 'original' homeland with a colonial periphery or frontier which characterized such different states as Spain (in the Americas), England (in North America), the United States (beyond the Eastern Seaboard), and Russia (in Siberia). In

Figure 5.1 Europe as a queen. This conveys both the symbolic sense of Europe's difference and its aristocratic superiority
Source: Munster, 1588.

each case, however, they had a colonizing experience closer to home. The Spanish had the south they had retaken from the Moors, the US coastal settlement was itself the result of invasion. The British Empire, for example, could be said to have begun in Ireland and elsewhere in the Celtic fringes of the British Isles (as the film *Braveheart*'s portrayal of English invasions of Scotland in the period 1290–1314 reminds its twentieth-century viewers). Colonialism started *in* Europe. It was later extended across the Atlantic and then to other world regions. In Ireland the rulers of England had their first major experience of subjugating, mapping, and renaming a land occupied by an (intermittently) hostile but (always) 'inferior' people. That this was incomplete because of persisting religious and political differences continued to concern the British élite down into the nineteenth century. The wild and defiant image of the Irish remained at odds with the civilizing mission of the English: to bring order and civilization to its oldest and closest colony. The first Others, therefore, were not abstract and unknown but concrete and familiar.

Beyond the limits of the homeland the *noblesse oblige* (the obligation of high rank to those lower down the social order) that aristocratic élites felt towards their social inferiors was projected more generally. At best the political ideas of most European imperialists were that, as in the case of the British in India,

> political power tended constantly to deposit itself in the hands of a natural aristocracy, that power so deposited was morally valid, and that it was not to be tamely surrendered before the claims of abstract democratic ideals, but was to be asserted and exercised with justice and mercy.
>
> (Stokes 1959, 69)

In this context, learning about and understanding the workings of colonized peoples was a vital part of the burden of spreading the light of European civilization into the world's darkest corners. The Nigerian novelist Chinua Achebe (1975, 5) captures the challenge and how it was thought of as follows:

> To the colonialist mind it was always of utmost importance to be able to say: I know my natives, a claim which implied two things at once: (a) that the native was really quite simple and (b) that understanding him and controlling him went hand in hand – understanding being a precondition for control and control constituting adequate proof of understanding.

Though obviously paternalistic and denigratory, this conception of others did incorporate a sense of wonder and even awe about the world into which Europeans had ventured. Indeed, many Europeans saw a nobility and simplicity in the lives and beliefs of the natives that they encountered. Analogies with ancient Greece and Rome often

Plate 5.1 Indians from Canada make a sacrifice to the Great Spirit. This drawing shows the wonder and simplicity (and classical analogies) that eighteenth century Europeans saw in primitive religion
Source: Bernard Picart, *Cérémonies et coûtumes réligieuses des peuples idolâtres*, Amsterdam, 1775. Getty Center, Resource Collections

underwrote this attitude, even down to the images used to represent religious rituals and performances (Plate 5.1).

Unlike the European empires proper, in which the European 'motherland' was separated from most of its colonies by large tracts of water, in the United States and Russia there was no such clear separation. Peripheries and frontier zones could be identified but there were not always obvious physical boundaries. The problem was resolved for a time in the Russian case by designating the Ural Mountains as the specific boundary between the European and Asian parts of the empire. Indeed,

> The basic geographical proposition that Russia divided cleanly and naturally into Asian and European sectors entered into the very foundation of the imperial ideology that was refined in the course of the eighteenth century. It was disseminated in the geography texts that began to appear in ever-greater numbers after 1750 and by the end of the century it had become a universally accepted truism.
>
> (Bassin 1991, 7–8)

In the United States, as European settlement spread into the continental interior during the early nineteenth century, the expanding frontier became the root and symbol of the filling out of national territory to the Pacific. Divine Providence was invoked initially to justify the invasion of the continent. The dominant justification, however, became the providential mission of 'America' to spread American ideals and institutions to the Pacific – and beyond. The rhetoric of mission peaked in the 1830s and 1840s when zealots wrote of American 'Manifest Destiny' and 'The Great Nation of Futurity.' In 1847 the US Secretary of Treasury placed in his annual report a section referring to the aid of a 'higher than any earthly power' which had guided American expansion in the past and which ' still guards and directs our destiny, impels us onward, and has selected our great and happy country as a model and ultimate centre of attraction for all nations of the world' (quoted in Pratt 1935, 343).

As this last quotation suggests, the agents of civilizational geopolitics, while invoking a general European outlook and agenda, came from different states which competed with one another as claimants to the mantle of the ancients. 'New' Romes abounded. Yet the claims rested more upon the imitation of French and English models of nation-building than upon some universal shared experience of the 'classical' European past. From successfully policing religious affiliations these prototype 'modern nation–states' had expanded their jurisdictions to construct a set of collective interests, outlooks, and traditions.

Particular dynastic and regional identities were reworked as 'national' cultural identities. Classical motifs were combined with local ones. The Napoleonic armies that swept across Europe in the aftermath of the French Revolution left architectural and organizational residues of its classical hubris; from the triumphal arches to the Code

Napoléon. But a second influence emanating from England and the German states soon spoke for a less remote past.

> Gothic churches, vaulted tombs, museums and houses of assembly, adorned with mementoes of medieval battles and national heroes, filled the lacunae of the nation's collective memory, and little children were taught to revere Arthur and Vercingetorix, Siegfried and Lemminkainen, Alexander Nevsky and Stefan Dusan as much, if not more than, Socrates, Cato and Brutus.
>
> (Smith 1991, 81)

The particular national genius was thereby highlighted against the achievements of a common but ancient past.

This harking back to medieval roots served to foster the sense of membership in singular communities with distinctive attributes that demanded 'self-realization' in 'historic homelands.' The French and English pioneers of nationhood, however imperfect or partial their own achievements, provided important role models for other Europeans, from the Greeks battling the Turks to the Germans and Italians struggling with their historic geographical divisions (described in Chapter 2). Not surprisingly it was to them, and particularly the Britain that emerged victorious from the Napoleonic Wars, that the new nation–states also turned for recognition of their status as modern nation–states.

A clear distinction was drawn between the possibility of this new statehood in Europe and its impossibility elsewhere. Within Europe (and wherever Europeans now lived) nations and states could be conjoined as limited territorial entities that balanced one another mechanically in a system of weights and balances. This analogy to Newton's mechanics provided the image of stability within which foreign policy between nominally equal states was possible. Outside this system was an unlimited space of primitive or decadent political forms that were candidates for conquest rather than recognition. The 'boundary' example of the Ottoman Empire provides a case in point. It was not recognized as a 'member' of the Concert of Europe until 1856 and even then the long-run disputes between it and Russia were never given the attention they deserved. The Otherness of the Turks was a fundamental barrier to their participation in a civilizational geopolitics that drew hard lines around its European homeland and even had trouble including such 'marginal' Europeans as the Russians and Americans.

NATURALIZED GEOPOLITICS

In anthropology the idea of *totemism* refers to the practice of interpreting nature in the vocabulary of human groups, such as kinship and genealogy, whereas *naturalization* does the opposite. It represents the human entirely in terms of natural processes and phenomena. This is what happened to geopolitics from the late nineteenth century down until the end of the Second World War, even though it had older roots (as indicated in previous chapters) and persisted in some respects thereafter (see the next section on 'Ideological Geopolitics'). Rather than a feature of civilization, geopolitics was now largely determined by the natural character of states that could be understood 'scientifically' akin to the new understanding of biological processes that also marked the period. The understanding of states and the geographical organization of the world economy that this gave rise to underpinned the inter-imperial rivalry of the geopolitical order from 1875–1945.

As the old Concert of Europe unraveled in the 1870s, two antagonistic groups of states emerged. One, led by Britain and France (with covert American support), stood for the coexistence of open trade and imperialism. The other, headed by Germany, was revisionist: concerned to build their own empires and challenge British financial dominance. By the 1890s this division had become the major feature of world politics. Its basis lay partly in the dramatic growth of the German economy but also partly in the intense nationalism of the period. National economy-making through the extension of national railway systems and the reorganization of economic space to accord greater centrality to national markets and finance was combined with the spread of literacy in vernacular languages and universal elementary education to produce an enhanced sense of national difference and exclusivity on the part of mass publics. Nationalism was no longer just for élites. In imitation of one another and as an outlet for investment during the economic depression of 1883–1896, a massive expansion of European empires brought much of the world that had remained outside the reach of European imperialism into a geographically specialized world economy. Different world regions took their place within a global division of labor, as plantation agriculture brought bananas to Central America, tea to Ceylon (now Sri Lanka) and rubber to Malaya (Plate 5.2). The main axis of capital accumulation lay in the home-colonies structure of the British Empire and the US–British relationship. The challenge for Germany and the other revisionists, such as Italy and Japan, was to build an alternative to this structure. Finally that proved impossible without taking on the British and their allies militarily. The climax arrived in 1914 when the alliance systems on either side of the inter-imperial rivalries set in motion in the 1870s came to life in response to the seemingly minor conflict over the political status of Bosnia-Herzegovina.

The apparently inexorable and inevitable slide into war associated with the First World War was the outcome of a mindset on all sides that saw war as the only way out of the political–economic impasse into which they had all slipped. The conduct of the war itself was also the outcome of a mentality in which the individual person had to sacrifice for the good of the greater whole: the nation–state that defined the totality of

Plate 5.2 'Jungles today are gold mines tomorrow.' Advertising poster of the Empire Marketing Board, September–October 1927
Source: Constantine, 1986

one's identity. Humanity had lost control of its destiny. Nature ruled in affairs of state. The naturalized geopolitics characteristic of this epoch had a number of tell-tale features: a world divided into imperial and colonized peoples, states with 'biological needs' for territory/resources and outlets for enterprise, a 'closed' world in which one state's political–economic success was at another's expense (relative ascent and decline), and a world of fixed geographical attributes and environmental conditions that had predictable effects on a state's global status.

Naturalization had a number of preconditions. One was the (apparent) separation of the scientific claim from the subject position of the particular writer or politician. Claims were made to universal knowledge that transcended any particular national, class, gender, or ethnic standpoint. So, even as a particular 'national interest' was self-evidently addressed, this was framed by a perspective that put it into the realm of nature rather than that of politics. A vital aspect of this, and another precondition for naturalization, was the conviction that observation of the world's political and economic divisions was a form of innocent perception from which generalizations about resources and power could be deduced. The pre-existing commitment to this view of perception and its associated mapping, naming, and citing of spaces as 'colonial,' 'European,' 'powerful' or 'backward' was not often acknowledged.

'Scientifically,' in the empiricist sense of the term, the world beyond the immediately familiar was blank and empty but was then filled and labeled according to its varying natural attributes as they appeared from Europe. The world was then known and *possessed* not just politically but epistemologically. This was the great achievement of naturalization; to have depoliticized inter-imperial rivalry into a set of natural and determining geographical 'facts of life.' The invention of political geography during this period is of one piece with this trend. On the one hand, the new field was openly touted by its advocates as a way of educating both élites and national populations in the ways in which physical geography constrained and directed state formation and empire building. On the other hand, it based its claims to seriousness on offering a perspective that was useful primarily for the nation–states from which its advocates came. Thus, Mackinder provided a global geopolitical model (see Chapter 1) but was concerned primarily about the implications of this model for the future of the British Empire.

There were, of course, conservative, socialist and anarchist currents of thought that tended to decry the merging dominion of naturalized modes of thought and action. But even they increasingly tended to flirt quite openly with claims to 'scientific' status that would put them beyond politics and into nature, so to speak. The voluntarist understandings of nationhood and class that had emanated from the American and French Revolutions were increasingly marginalized. For example, the polemic of Ernest Renan (1882) against the German 'blood and soil' conception of nation on behalf of a voluntarist one (of choice or will) was rapidly eclipsed.

One of the most important elements in naturalized geopolitics was the distinction between imperial and colonized peoples. Citizenship laws, classifications of colonies (particularly between European settler colonies and others), religious missionary activities, and even the emerging division of labor between university subjects (e.g. sociology versus anthropology), operated on this distinction. But this distinction was not merely a 'recognition of necessity;' a pragmatic response to an obvious spatial division. Ironically, it rested on the view that (some) Europeans had become masters *of* nature as a result of superior 'fitness' in a natural process of evolution. Transforming or displacing more primitive peoples was justified because as Charles Darwin (1839, 520), one of the intellectual sources for this type of reasoning, had put it: 'the varieties of man seem to act on each other in the same way as different species of animals – the stronger always extirpating the weak.'

In the United States in the 1880s scientific racial claims largely replaced providential/civilizational ones as the basis for American destiny. Anglo-Saxons (and other 'Teutons') were identified as exemplars of superior evolutionary fitness. Their superior political principles, growth in numbers, and economic power were evidence enough (notwithstanding contradictory evidence in the Census of the United States which suggested the greater growth of other groups). One leading proponent of the fitness argument saw an Anglo-Saxon future. 'The day is at hand,' he wrote, 'when four-fifths of the human race will trace its pedigree to English forefathers, as four-fifths of the white people of the United States trace their pedigree today [1885]' (quoted in Pratt 1935, 348).

The principle of natural selection thus filtered down from scientific theorizing into popular culture largely in terms of the idea of the 'survival of the fittest.' This became a staple of the journalistic accounts and travel-writing that accompanied the great explosion of European imperialism at the turn of the century. An older idiom justifying colonialism in terms of moral uplift and religious grace gave way to a discourse of racial competition and dominion. Initially invoked to distinguish the major 'races' of humanity, it quickly lent itself to more refined distinctions, as with the American 'Anglo-Saxons,' and served as an important inspiration for the racist ideologies – racial anti-Semitism, anti-Slavism – and eugenics programs that flourished politically in Europe and North America in the 1920s and 1930s. These ideas were shared internationally rather than associated singularly with any one particular country such as Germany. For example, to recall the example of England and Ireland from the previous section on civilizational geopolitics, the late nineteenth century saw a racialization of the presumed inferiority of the Irish. To the historian Lord Acton the whole language of English success and Irish failure as imperialists was attributable to the natural inferiority of the Irish race:

> The Celts are not among the progressive . . . races but those which supply the materials rather than the impulse of history, and are either stationary or retrogressive . . . They are a negative element in the world . . . and waited for a foreign influence [the English] to set in action the rich treasure which in their own hands could be of no avail.
>
> (quoted in Curtis 1983, p. 53)

Within Europe, Jews were especially vulnerable to the growth in scientific racism. Religious anti-Semitism had deep roots in Europe going back to the Middle Ages. But not only did the nineteenth century bring rapid advancement to Jews as a group as they moved out of ghettos into society, but some non-Jewish groups 'experienced rootlessness, fragmentation, and estrangement from a once securely anchored, familiar world' (Barnouw 1990, 79). As a result, Jews became objects of hatred and fear from those nostalgic for a world they had lost but without good prospects in the new national–industrial society. Demagogues readily portrayed Jews as rootless cosmopolitans in a

Plate 5.3 'Behind the enemy powers – the Jew.' Nazi racial propaganda
Source: Bildarchiv Preussischer Kulturbesitz, Berlin

European world dividing into parochial nation–states (Plate 5.3). There was no longer a space for difference within the boundaries of states. Jews were dangerous polluters of national homogeneity. Every 'race' was seen as requiring its proper place. Jews were increasingly portrayed in racial terms. In the work of the nineteenth-century political geographer Friedrich Ratzel (1844–1904), the founder of the ecological theory of race, the Jews are seen as the one race most 'out of place.'

> In the Near East they were productive (for example, creating monotheism) but in Europe they have no real cultural meaning. The association of place and race is linked in the rationale of the German in Africa or the Jew in Europe. They are presented as mirror images, for while the German in Africa 'heals,' the Jew in Europe 'infects'.
>
> (Gilman 1992, 183–4)

The Nazi geopoliticians of the 1930s came up with formalized schemes for combining imperial and colonized peoples within what they called 'pan-regions.' However fanciful in terms of possibilities of overcoming the political–economic relationships of the time, such mappings did express in extreme form the common assumption that the world was constituted by racial groupings that could be neatly divided into two 'types' of peoples. The one essentially existed to serve the other. Dominant and subordinate races were joined together territorially in the pan-regions. Whatever the precise influence of the Nazi geopoliticians, particularly Karl Haushofer, upon the practices of the Nazi regime, there is little doubt that their ideas fit into a larger political context in which notions of racial hierarchy were conjoined with conceptions of state 'vitality' to justify territorial expansion. The *Zeitgeist* (spirit of the age) even drew into its orbit intellectuals with impeccable anti-naturalist credentials. The classic case in point is the German philosopher Heidegger (1959, 39) who wrote in 1935, apparently without tongue in cheek:

> We are caught in a pincers. Situated in the center, our *Volk* incurs the severest pressure. It is the *Volk* with the most neighbors and hence the most endangered. With all this, it is the most metaphysical of nations.

At the same time, either through analogy or literally, the European territorial state acquired a status as an organism with its own 'needs' and 'demands.' This too was a transposition from evolutionary biology. But it had older roots. German idealist philosophers of the early nineteenth century such as Hegel (1770–1831) and Fichte (1762–1814) had regarded the state as a being or entity with a life of its own. It was but a short step from this to the idea of the state as an organism, a step facilitated by the spread of biological ideas into the emerging social sciences and the rhetoric of politicians in the 1890s.

Like all organisms a state must struggle against the environment (in this case, other legitimate states and 'empty spaces') to survive. This required that it acquire space and resources to feed its healthy growth. The rebirth of militarism in the late nineteenth century after its decline in the aftermath of Napoleon was a corollary trend that further fed the image of the state as a permanently embattled entity that could prosper only if individuals and classes subordinated their particular interests to the interest of the larger whole. The doctrine gained expression in terms of three central points: the harmony of state and nation, natural political boundaries, and economic nationalism.

The late nineteenth century was a time of tremendous social disruption in Europe. Not only were there massive movements of people within Europe and across the oceans leading to an explosion of urbanization, but increasing capital mobility also undermined the national and local circuits of savings and investment that had previously given an appearance of long-term commitment to the activities of capitalists. These trends were particularly strong in Britain where by the 1890s a reaction had set in against their social consequences. One was the growth of political movements pushing for political and economic rights for workers and women. Another, increasingly popular with political élites, was directed at nipping these new movements in the bud by 'turning the clock back' to a time when all social groups were thought to have lived in local social harmony. Only this time the social harmony was to be realized in the conjunction of the nation with the territorial state.

A mythic medieval community in which each social stratum knew its place and duty to the whole was projected onto the nation as a whole. It was this recreation of the local past in the national present that served to give the state its organic character. An important contrast was drawn between this and the workings of a *laissez-faire* economy. Such an economy was seen as undermining the organic unity of the nation–state. To some influential conservatives it was seen (and still is seen!) as a source of decay, what Oswald Spengler (1926) was to term after the First World War *The Decline of the West*. The 'founders' of disciplinary geopolitics, the Swede Kjellen and the Englishman Mackinder, both subscribed to this viewpoint.

This perspective was widely shared by emerging political élites throughout Europe (and its overseas extensions). In Russia the Urals were increasingly discounted as a geographical divide. One influential intellectual current came to see 'Russia as a transcendental geo-historical, geo-political, geo-cultural, geo-ethnographical, and even geo-economic entity, designated by a new generic term: *mestorazvitie*' (Bassin 1991, 16). In Germany local patriotism or identity with the *Heimat* (or homeland) was channeled by conservative political parties into an overarching identity with the *Reich*. Only through the advance of the *Reich* could the *Heimat* be defended. In the United States, until the Civil War, 'the United States' was typically a plural noun. Afterwards, it became singular. This transformation is symbolically associated with President Lincoln's famous

Gettysburg Address of 1863 when: 'What had been a mere theory of lawyers like James Wilson, Joseph Story, and Daniel Webster – that the nation preceded the states, in time and importance – now became a lived reality of the American tradition' (Wills 1992, 145).

Another element in the view of the state as an organic entity was the idea that a state had 'natural boundaries.' This implied, first of all, that the historical boundaries were not necessarily the proper ones. The territorial *status quo* of the Concert was now called into question. But it also implied that all of the members of a putative nation or ethnic group had a natural right to live within the boundaries of the state. Finally, it also opened up the possibility of using natural features to designate the natural area of the state. Swedish conservatives, such as Kjellen, argued against Norway's independence partly because they claimed that the Scandinavian mountains were not a natural boundary. The Nazi concept of *Lebensraum* (borrowed from the German geographer Ratzel), justifying German territorial expansion in *Mitteleuropa* (and elsewhere) and what was actually called the 'intellectual liquidation' of Poland (Burleigh 1988, 50), had their roots in the notion of natural boundaries.

This logic was never extended to the 'stateless' colonized world. This is most clearly illustrated in the case of Africa. In the wake of the exploration of Africa, conquest and colonial rule came quickly and devastatingly. In 1884–85 the major European powers agreed at the Congress of Berlin to stake out their spheres of influence in Africa. The 'Scramble for Africa' which followed over the next twenty years produced lines on a map which had little relation to underlying cultural or economic patterns. Elsewhere, the establishment of colonial boundaries was less hasty if often no less arbitrary. These designations continue to haunt these regions to this day.

Lastly, a major tenet of 'organic conservatism' was economic nationalism. The state was seen as defining the basic unit for economic transactions. Firms and individuals were held to be subordinate to the greater needs of the nation–state. This too was biological in nature. Writers as different in other respects as the English economist Hobson (who influenced Lenin's thinking) and the English geographer–politician Mackinder both shared organic definitions of national interest as the driving force behind economic growth. To Hobson empire sacrificed the 'home' economy, while for Mackinder empire was the means of maintaining the economic basis to the military power that was essential for national survival. What they and others would have agreed on was the organic unity of the national economy as a 'going concern.' The German *Weltpolitik* (roughly, economic imperialism), though often at loggerheads with doctrines of *Lebensraum* or territorial expansion, nevertheless provided a similar popular logic for acquiring overseas markets and sources of raw materials to underpin German industrial success.

The idea of a 'closed world' was vital to the plausibility of the language of biology for understanding the European state as an organic entity. As frontiers 'closed' in North

America and the world's land masses were incorporated into the world economy, control over territory appeared to be a crucial prerequisite for economic growth. For British élites in particular the certainties of a world economy that worked in their favor seemed to be gone. They were faced both by protectionist rivals and a domestic commitment to free trade that the new geopolitical order had made anachronistic. 'The British economic lead had evaporated, and for the British the world did indeed seem to be shrinking, to be closing in' (Kearns 1993, 29).

But there was more to it than this. From 1880 to 1914 'a series of sweeping changes in technology and culture created distinctive new modes of thinking about and experiencing time and space' (Kern 1983, 23). Such innovations as the telegraph, the telephone, the automobile, the cinema, the radio, and the assembly line compressed distance, truncated time, and threatened social hierarchies. The global spread of railways and the invention of the airplane were perhaps the most important challenges to conventional thinking about time and space. The sense of a closed world, therefore, was neither illusory nor the product of a uniquely British sensibility.

In a 'closed world' a premium would be placed on relative national efficiency. States must therefore organize themselves to increase their productivity relative to their rivals. No state could afford to rest on its laurels.

> Nationalism and protectionism helped countries mobilize the resources of their earth and their people . . . A country without access to the full complement of modern industries was vulnerable, would be a pushover in a war, and thus would attract the bellicose attention of more well-balanced nations.
>
> (Kearns 1993, 18–19)

The major political dispute in late nineteenth-century/early twentieth-century Britain concerned the merits of maintaining the policy of unilateral free trade introduced in 1870. Both sides adopted the language of national interest and 'shared a belief in the importance of "national economic power" but they lacked agreement on exactly what the concept meant or how it should be measured' (Friedberg 1988, 79).

If the liberal economics of the nineteenth century came under attack from the 1890s on, it suffered a nearly fatal final blow from the Great Depression of the 1930s. The mass unemployment of the time produced a variety of intellectual and political reactions. State corporatism pushed the organic analogy to its limits. From this point of view, ascendant in fascist Italy, Nazi Germany, Spain, and Portugal, old economic assumptions no longer applied. The national unit was now pre-eminent. State corporatism was 'a system of interest and/or attitude representation, a particular modal or ideal–typical institutional arrangement for linking the associationally organized interests of civil society with the decisional structures of the state' (Schmitter 1971, 86). The state could now be freed from guaranteeing such values as individual liberty and equality to pursue its 'own' agenda of security and foreign affairs with the total support of its economy.

The breakdown of the capitalist world economy also offered an opportunity for some form of socialist internationalism. But Stalin's adoption of 'socialism in one country' in a Soviet 'socialist motherland' effectively reproduced the territorial economism characteristic of other regimes rather than called it into question. This 'model' of socialism – central economic planning, collective agriculture, and new political élite – did threaten established élites, however, particularly where communist political parties affiliated with the USSR were strong, and led to the association between socialism and 'subversion' that was to be so important in geopolitical discourse after the Second World War.

It was the rationale for government management of the economy provided by the English economist Keynes that provided the most important justification for stimulating the national economy in Britain and the United States in the 1930s and 1940s. Liberalism remained important in both of these countries when it had long disappeared elsewhere. Keynes provided a way of 'squaring the circle', by arguing for a synthesis of private and state activity through 'countercyclical demand management.' The national economy, however, was the basic unit for Keynesian macroeconomics just as it was for the other political–economic philosophies. Indeed, its basic measures such as national 'propensity to save,' national investment, and national productivity were indicators of national efficiency that earlier and more illiberal figures would have readily recognized for what they were.

The final feature of naturalized geopolitics was its emphasis on the determining character of geographical location or environmental conditions. The relative success of different states in international competition was put down to absolute advantages of location and to superior environmental conditions. 'Marchland' states (states on the edges of land masses) were seen by military strategists as possessing intrinsic advantages over 'interior' states because they had fewer contiguous or neighboring states and, hence, fewer potential adversaries. 'Maritime' or sea-power states were seen as out-flanking 'continental' or land-power states in control over the oceans, the main means of global movement. Only the coming of the railway had called this into question; and this because of the relative size or weight of the Eurasian land mass (or 'heartland') in relation to the difficulty of coalescing and policing the 'insular crescent' (Mackinder) or 'rimland' (Spykman) around its edge. The use of the Mercator projection to represent the earth's land areas served to exaggerate the sense of a world dominated by Eurasia (especially Siberia) because of its systematic enlargement of polarward areas relative to equatorial ones (see Figure 1.3 in Chapter 1).

This spatial–geographical determinism associated with formal geopolitical models, however, was never as popular as a less specific (and more ambiguous) environmental determinism. From this point of view the Great Power potential of states was a function of their industrial prospects which, in turn, could be traced to their natural resources (particularly energy resources) and ability to exploit them. Some went further and claimed that this ability was itself 'determined' by climate. From 1900 until the 1930s such views were not exceptional. Indeed, they formed the mainstream of opinion, particularly among those educated in such academic fields as geography, geology and biology. Much of the academic geography of the period in Germany and the English-speaking world involved elaborating systems of environmental/geographical accounting; classifying states and regions in terms of inventories of resources, racial characteristics, economic and political organization, and climatic types. These were taught in schools and became the 'conventional wisdom' about why some places had 'developed' while others had lagged behind. The manuals issued to US soldiers during the Second World War are classics of this genre; their title is revealing of their content: *Geographical Foundations of National Power* (US Army 1944): natural attributes determined national destiny.

The Second World War, therefore, was not external to this geopolitical discourse. Naturalized geopolitics pervaded the representations of the war itself in the film and cartographic propaganda that both sides engaged in. On the Axis (Germany–Italy–Japan) side it provided the logic for the war (expand or perish) and (in the German case) for the anti-Semitic Holocaust that was one of its central features. On the Allied side the war was seen as a struggle for survival by peaceful, maritime states whose success was based on their capacities for invention and trade. The outcome of the war brought to an end the geopolitical order of inter-imperial rivalry and created the conditions for the construction of a new postwar geopolitical order: 'The emerging order was a qualitatively distinctive one, characterized by the breakup of the old colonial empires through the decolonization process, and the emergence of the United States as an economic, military, and political hegemon' (Biersteker 1993, 16).

IDEOLOGICAL GEOPOLITICS

By definition the modern geopolitical imagination is ideological, if ideology is defined as an amalgam of ideas, symbols, and strategies for promoting or changing a social and cultural order or, as the anthropologist Paul Friedrich (1989, 301) puts it, 'political ideas in action.' After the Second World War, however, the geopolitical imagination was centered much more explicitly around competing conceptions of how best to organize the international political economy. Cold War geopolitics was 'linguacultural' (Friedrich's term) more than civilizational or naturalized. By this I mean that the values, myths, and catchwords drawn from the experiences of the two victorious states, the

United States and the Soviet Union, were to define and determine the terms of the geopolitical imagination of the period. One of these, the United States, was to prove the more effective in gaining widespread acceptance for its 'model' of political–economic organization. But its success relied heavily on the active presence of the other as a point of comparison and threat.

Under the dictatorship of Stalin from 1943 to 1947 the Soviet Union built a formidable military economy that required as its premise the existence of a major external threat. The recent invasion by Nazi Germany meant that a sense of threat was not difficult to sell to the Soviet population. Associated with this sense of external danger was an identification with regimes and revolutionary movements that were opposed to the American/Western model of economic development. On the American side the US government set out from 1944–47 onwards to sponsor an international order of economic liberalism in which military expenditures would provide a protective apparatus for increased international trade. This would in turn redound to the advantage of American businesses and stabilize the American economy. The key institutions and rules spread rapidly in the immediate postwar period. All of the major industrialized countries except the Soviet Union accepted the rules of the international economic game as written in Washington, DC, either through external inducement (as with the aid from the 1947 Marshall Plan) and coercion (as with the British dollar loan of 1946) or through direct intervention and reconstruction under American auspices (as in West Germany and Japan). Many of the elements of this system have lasted into the 1990s, although some faded away earlier. They are:

1 indirect stimulation of economic growth by means of fiscal and monetary policies;
2 commitment to a growing global marketplace based on a global division of labor;
3 accepting the dollar as the principal world currency;
4 hostility to Soviet-style economic planning;
5 assuming the burden of policing political changes that could be construed as damaging to the stability of the world economy.

In the end domestic conditions in the two main states undermined the global status of each. By the 1980s the Soviet economy had failed to deliver both improved military equipment and improved living standards for the Soviet population as the Communist Party and its leadership sank into complacency. For the United States, internationalization was too successful. By the early 1970s the growth of other economies produced pressures on the dollar and the drain of policing proved increasingly expensive.

Not simply international relations but also the economic and political arrangements of all of the world's states beyond the Soviet sphere of influence (China for a few years in the early 1950s, Eastern Europe until 1989) came under the spell of the American system. The vital glue for this system was provided by the political–military conflict with the Soviet Union. Even during times of coexistence or *détente* the overarching Cold War conflict served both to tie such important states as Germany and Japan into the US system and to define two geopolitical spheres of influence in which each generally accepted the other's dominance. This imposed a stability on world politics, since the United States and the Soviet Union were the two principal holders of nuclear weapons and they could threaten to escalate conflicts if their 'fundamental strategic interests' were threatened, even as it promoted numerous 'limited' wars in the Third World of former colonies where each armed surrogates or intervened themselves to prevent the other from achieving a successful 'conversion.' For all of their political and economic weaknesses, however, many Third World countries had to be wooed and cajoled. Unlike in the previous period, the world map was no longer a vacuum waiting to be filled by a small number of competing Great Powers. The United States and the Soviet Union remained attached to anti-colonial philosophies even as they engaged in imperial competition. They had to win friends and influence people to extend their spheres of influence; they could no longer simply impose external rule.

In this setting, ideological geopolitics developed the following major characteristics: a central systemic–ideological conflict over political–economic organization; 'three worlds' of development in which American and Soviet spheres of influence vied for expansion into a 'Third World' of former colonies and 'non-aligned' states; an homogenization of global space into 'friendly' and 'threatening' blocs in which universal models of capitalism–liberal democracy and communism reigned free of geographical contingency; and the naturalization of the ideological conflict by such key concepts as containment, domino effects, and hegemonic stability. Older themes from civilizational and naturalized discourses were worked into the new discursive space. The two most important have been the backward vs. modern polarity and the idea of 'national security.' These will be mentioned as need arises.

The Cold War began as a series of US policies designed to rebuild Western Europe after the Second World War but became a system of power relations and ideological representations in which each 'side' defined itself relative to the other that it was not. This happened in an *ad hoc* manner but had deep linguacultural roots.

The United States and the Soviet Union had been allies during the war. In the aftermath of the war a basic question concerned how Western Europe was to be organized politically and economically. US moves to spread American views of how this might best be done met with resistance from the Soviet Union and its allied communist parties in Western Europe. From 1945 to 1950 this initial conflict was

> reinforced by a crisis here, a Soviet move there, and an analysis of the protagonists [in the United States] which insisted that Moscow was impelled to expand and that only the United States could prevent it from achieving world domination.
>
> (Cox 1990, 30)

The 'loss' of China to Communist Revolution was a particularly important element in deepening US–Soviet mutual hostility.

But American and Soviet mutual suspicion had origins that went back to the Bolshevik Revolution of 1917. There had been a 'red scare' in the United States then that raised the specter of domestic subversion. US military forces also intervened in the Russian Civil War against the Bolsheviks; the US government knew what it was up against. In the 1920s a fear that 'alien' influences threatened the workings of America's unique institutions led to a resurgence of isolationism. 'Foreigners' were dangerous. The United States refused to recognize the new Soviet government in the old Russian Empire.

Both the United States and the Soviet Union were peculiar states. They both had origins in revolutions with explicit ideological agendas. They both claimed popular mandates that transcended particular ethnic, class, or regional interests. They both offered themselves as uplifting lessons in political–economic experimentation to a world where cynicism was rampant. In this context, the practices of 'foreign policy' took on a special importance. In two states where what precisely was either 'American' or 'Soviet' was unclear, the threat of things 'un-American' or 'anti-Soviet' became central to national identity.

Ideologically, neither saw itself as a limited territorial entity, though in each there had been long-standing disputes over the geographical scope of their respective revolutions. 'Isolationism' versus 'internationalism' has been the typical language of foreign-policy debate in the United States. But by the 1940s the dominant élites of each aspired to global ideological dominion. In the Soviet case this was obvious in the ideological lineage that informed official ideology (Marx – Engels – Lenin), if less obvious in the official practice of isolating the Soviet economy from external linkages. In the United States the urge to export the American 'ethos' was long-standing. Certainly, from the 1890s onwards 'America' itself was increasingly seen by leading Americans as an idea that could be sold:

> American traders would bring better products to greater numbers of people; American investors would assist in the development of native potentialities; American reformers – missionaries and philanthropists – would eradicate barbarous cultures and generate international understanding; American mass culture, bringing entertainment and information to the masses, would homogenize tasks and break down class and geographical barriers. A world open to the benevolence of American influence seemed a world on the path of progress. The three pillars – unrestricted trade and investment, free enterprise, and free flow of cultural exchange – became the intellectual rationale for American expansion.
>
> (Rosenberg 1982, 37)

This outlook had more idealistic origins. In 1823 President James Monroe had issued the statement that the United States would oppose all foreign intervention in the American hemisphere. This doctrine became as sacred to many Americans as the Declaration of Independence and the Constitution. Over the years, however, it served to justify US military intervention anywhere in Latin America and the Caribbean, even though this was to give it a meaning quite different from the original sentiment expressed by Monroe. During the Cold War period the Monroe Doctrine licensed US governments to intervene throughout the Western Hemisphere and, by extension, into the rest of the world. An originally democratic impulse was thus recycled under new conditions and in so doing undermined what it had originally stood for.

Ironically, given the global pretensions of each, domestic subversion by foreign agents is a dominant theme in each state's view of its own history, as is an assumed vulnerability to external threats. Again, the perpetual threat to revolutionary achievements is probably of some significance. In the United States appeals to public opinion have also been of importance in legitimizing any kind of foreign policy. Hence, exaggerating vulnerability and invoking the need for 'self-defense' became an important means of mobilizing public opinion on behalf of foreign ventures. The arrival of nuclear weapons (and long-distance delivery systems) led to an American sense of connection to events elsewhere in the world, through the possibility of escalation of local conflicts into a global nuclear one, that previously had been missing. The long history of American territorial security has, however, perhaps left people without much of a basis for judgment about external 'threats,' even as the Soviet experience of invasion and mass murder as recently as the Second World War provided a more obvious basis for collective paranoia.

One important consequence of this shared sense of vulnerability was an idealization by each of the other. Each became a super-potent adversary in the eyes of the other. This is particularly obvious in the American case where there was systematic exaggeration of Soviet economic and military capabilities. For example, Holzman (1989) has shown systematic exaggeration in official US estimates of Soviet military spending from the 1960s to the 1980s. Gervasi (1988) in his annotated version of the Pentagon publication, *Soviet Military Power*, finds evidence of massive over-statement of Soviet military capabilities that we now know, after the collapse of the Soviet Union, to be absolutely on the mark. As late as 1988 US foreign policy was still based on the continuing existence of a singular Soviet threat to the US position in the international political economy. Even as the Soviet Union disintegrated in 1988–89 a group of leading US experts on military and foreign affairs produced a report that assumed an indefinite continuation of Cold War bipolarity. Though the economic problems of the United States – relating to huge trade and federal government deficits – were well known, it ignores them. Instead, it engages in statistical and cartographic legerdemain to demonstrate *increased* American vulnerability to the Soviet threat. It is no exaggeration to say that political élites in each country became so obsessed with each other that, as one of their number put it, he hoped they would not appear to future historians as 'two dinosaurs circling one another in the sands of nuclear confrontation' (Gorbachev 1987–88, 494).

In the context of peculiar 'revolutionary' states with quite different recipes for political–economic organization, abstract terms such as 'communism' and 'capitalism' took on culturally loaded meanings. As the United States came to personify capitalism, so did the Soviet Union represent communism. Each state became the geographical manifestation of an abstract political economy. Each was foreign and dangerous to the other. The constitutive divide in world affairs after the Second World War thus had specific linguacultural roots that reduced communication to a repetition of buzzwords about the nature of the other party.

Joanne Sharp (1993; 1996) has shown that in the American case after the end of the US–Soviet wartime alliance a widely circulating magazine like the *Reader's Digest* served up a steady diet of articles by 'experts' that created a 'common-sense' knowledge about the Soviet Union and the threat it posed to Americanness. Its readers were not unprepared. The *Digest* had been worried about the Soviet Union since the early 1920s. Now readers were bombarded with accounts of the total difference with and opposing character of the Soviet Union to the 'actually existing' United States, not simply differences in aspiration or historical experience. Each was masqueraded as the actual opposite of the other. Such representations had powerful effects in mobilizing public opinion into the Cold War consensus that progressively engulfed American politics from 1947 until the Vietnam War of the late 1960s.

Certain domestic interests were served by this geopolitical reductionism. Once under way this reinforced its pervasiveness. In the Soviet Union it disciplined potential dissidents into support of a monolithic state apparatus. In the United States it produced a consensus around a 'politics of growth,' an enlarged military economy, and opposition to any politics (usually construed as 'socialist' or leftist because of associations between socialism and the Soviet Union) that would subvert the country from within. In other words, it eroded the possibility of an open, competitive democratic politics. The very identity of being American became associated with a narrow political spectrum at home and a virulently anti-socialist/anti-Soviet position (they were rarely distinguished) abroad. 'The simple story of a great struggle between a democratic "West" against a formidable and expansionist "East"' became 'the most influential and durable geo-political script of [the Cold War] period' (O Tuathail and Agnew 1992, 190).

As the 'leader' of the West the American President played a key role in giving meaning to the Cold War.

> In ethnographic terms, the U.S. President is the chief *bricoleur* of American political life, a combination of storyteller and tribal shaman. One of the great powers of the Presidency,

> invested by the sanctity, history and rituals associated with the institution – the fact that the media take their primary discursive clues from the White House – is the power to describe, represent, interpret and appropriate.
>
> (ibid., 195–6)

Through the recycling and repetition of certain images and themes taken from the American past, the President can give a sense of continuity to the practice of foreign policy.

On 12 March 1947 President Truman laid an important piece in the groundwork for the Cold War geopolitical imagination in a speech to the US Congress that while it 'drew the line' against communism in Greece also integrated this into American history by referring to 'free peoples who are resisting attempted subjugation by armed minorities or by outside pressure.' An American audience knew well that this was a reference to their own experience. Connecting the Cold War with the American Revolution was thereafter a central theme of Cold War discourse. President Reagan, for example, was to regard the Nicaraguan *Contras* ('anti-Communists') of the 1980s as the 'moral equivalents of the [US] founding fathers.'

After the Second World War the US political élite read three lessons from the war and the Great Depression that preceded it: that appeasement of potential aggressors was dangerous, that American national security depended upon the global balance of military forces and not just the balance in the 'Western hemisphere,' and that the United States must everywhere oppose attempts at building regional economic blocs or 'pan-regions' such as had emerged in the 1930s. These became the persisting themes of Cold War discourse as they were incorporated into US government practice.

They led inevitably to a 'forward posture' in confronting the Soviet Union. Europe was the main setting for challenging the possibility of Soviet expansion beyond the confines accepted by Roosevelt, Stalin, and Churchill at the Yalta Conference in early February 1945. Both the United States and the Soviet Union came to share a mutual interest in this scenario. Berlin and its division became the symbolic center of the Cold War, where each side confronted and stared down the other (Plate 5.4). The geographical division of Europe guaranteed US and Soviet status as the dominant partners in the alliances they formed in the two halves of Europe. Very quickly, the common goal became the maintenance of balance. 'In this sense the Cold War was more of a carefully controlled game with commonly agreed rules than a contest where there could be clear winners and losers' (Cox 1990, 31).

Although there were episodic uprisings and movements against this spatial division, particularly in Eastern Europe, but also on the part of the Gaullist movement in France, the 'hottest' consequences of the Cold War were felt largely in the 'Third World.' This term is itself a product of the Cold War. Although it was first coined in France in the late

Plate 5.4 Dismantling the Berlin Wall: symbol of the Cold War after it was erected by the East German government in 1961 to physically divide their section of the city from the Western/capitalist West Berlin, its fall on the night of 9–10 November 1989 represented the symbolic end of the Cold War. Here people celebrate the end of the division of Berlin on the night in question at the Brandenburg Gate, the historic center of the city.

1940s to refer to a possible Third Estate or Third Way, it soon began to refer to those parts of the world outside the settled spheres of influence of the 'superpowers,' the First and Second Worlds, in which their global conflict would be concentrated (see Chapter 2). This logic was to prove visionary. Geopolitical space was conceptualized in terms of a threefold partition of the world that relied upon the old distinction between traditional and modern and a new one between ideological and free. Actual places became meaningful as they were slotted into these geopolitical categories, regardless of their particular qualities.

The threefold categorization turns on a combination of the essential attributes produced by a cross-classification of the pairs of terms 'traditional (backward) vs. modern' and 'ideological vs. free.' At a first cut, the modern 'developed' world is distinguished from the traditional 'underdeveloped' world (the Third World). At a second cut, the modern world is divided into two parts: a non-ideological (capitalist) or natural (free) First World and an ideological (socialist) Second World. Of course, this is the dominant American rendition of who is and who is not 'ideological.' In the context of the Cold War only left-wing or socialist politics received this attribution.

The classification has not been contained by its particular origins. Indeed, the concept of the Third World has come to signify resistance to the discursive domination of the superpowers and the possibility of alternative paths to development. For example, theories of development based on the idea of 'dependency' and emanating from such Third World settings as Latin America and Africa, reject the argument that the sources of uneven development are to be found within poorer countries themselves; they lie in their external connections to the world economy. The Non-Aligned Movement of such Third World leaders as Tito (Yugoslavia), Nehru (India) and Nasser (Egypt) attempted to subvert the very idea of a simple choice between the United States and the Soviet Union. For dominant political and academic élites on the two defining sides at the height of the Cold War, however, there was little doubt that subordination and imitation were the order of the day. The Third World was that vast geographical zone not yet committed to a particular path to modernity. The success of the superpowers would lie in their relative abilities to recruit candidates for their respective models of political economy from the ranks of the Third World. At the same time, particularly for the Americans, there was the need to counter governments that came to power without a clear commitment to US policies. From Guatemala in 1954 to Granada in 1986 one or other of these rationales was provided for American foreign intervention. The USSR took a similar line in Hungary, Czechoslovakia and Afghanistan.

In each case the local situation was invariably tied to the larger global context. This was because no particular place had singular attributes, only characteristics that followed from its position in the abstract spaces of the Cold War. It was 'friendly' or 'threatening,' 'ours' or 'theirs.' This homogenization of global space made knowing the details of local geography unimportant or 'trivial.' All one needed to know was: whose side are they on?

Identifying or defining places without a larger classificatory grid is not possible. All maps depend on projections, for example. But 'local knowledge' is possible to a degree that Cold War geopolitics essentially denied. So-called area specialists in government and university departments were frequently frustrated by their lack of influence. Certainly, in the United States the 'big picture experts' like Henry Kissinger, Alexander Haig and Zbigniew Brzezinski were always more important than the area specialists in policy disputes. Local knowledge was 'background' that could be called on when needed. It did not determine the real identity of places. That was done by putting them into the global frame of reference.

Conflicts with apparent local roots were thus read as local manifestations of the superordinate global one. Links to outside powers in the form of supplies of arms or the furnishing of advisors were read as the only causes of local conflicts. States following their own development or foreign policies were seen as candidates for neutralism or 'Finlandization;' a dreaded condition involving trading with the Enemy but without unequivocal commitment to one side or the other. States were not autonomous actors but agents for one side or the other, expected to fulfill the political and economic goals of the superpowers. Formal organizations such as the North Atlantic Treaty Organization (NATO), the International Monetary Fund (IMF), and the World Bank on the American side and the Warsaw Pact and the Council for Mutual Economic Assistance (CMEA) on the Soviet side institutionalized the division. In the various United Nations organizations (such as the Security Council and the General Assembly), created at the end of the Second World War to inaugurate a more pacific world order, the two sides expressed their mutual contempt and hostility. The Third World was given a voice but precious little else in UNCTAD (the United Nations Conference on Trade and Development).

The global spatial division recapitulated on a world scale a tendency strong in both American and Soviet (Russian) cultural history to draw strong lines between the space of the 'Self' and the space of the 'Other.' In both cases, if best known in the American case, political discourse has been shaped around the experience of the *internal frontier* which during national development had separated civilization from savagery, domesticated from wild, good from evil. The entire course of American history is read in many school history texts as a realization of Manifest Destiny: that Americans are a chosen people destined to expand in territory, wealth and influence. Similarly, in the Soviet case, Russians long saw themselves as the bringers of light into the cultural darkness of Siberia and Central Asia. The closeness of the colonial frontier allowed each to think of itself as engaged in something quite different from the nasty imperialism of the real Europeans.

Another incentive to a neat bifurcation of global space came from the possession of large numbers of nuclear weapons by the superpowers. If allies became involved in conflicts there was always the danger of escalation from local/conventional to global/ nuclear war as they drew in their respective 'senior partners.' This served to discipline allies since they provided the places where escalation (i.e. nuclear war) would commence. 'Bipolarity,' in the sense of two overwhelming military powers, thus imposed a balance of terror on the world as a whole and froze political boundaries as they were in 1945. But it also restrained the superpowers and, by the 1960s, gave leverage to allies who could exploit the superpowers' fear of mutual destruction once they were able to threaten one another directly with arrays of ICBMs. Nuclear weapons are not inherently stabilizing even under conditions of bipolarity because there is always a danger of automatic escalation. However, the concentration of nuclear ownership certainly served to highlight the distinctiveness of the two superpowers, whatever their other military deficiencies.

George Kennan, a US official in the Soviet Union at the end of the Second World War, in his famous 'Long Telegram' from Moscow and 'Mr. X' article in *Foreign Affairs* in July 1947, argued that the Soviet Union was a totally alien space with which there could not be meaningful compromise. This claim had an important effect on the policy choices that the Truman administration made relative to the Greek Civil War (1947) and the founding of NATO (1949). Little surprise, then, that the image of huge blocs of space without meaningful internal variation became a fundamental part of the Cold War geopolitical imagination.

Three geopolitical concepts played especially important roles in naturalizing these understandings of space and global politics for Americans and others. These were containment, domino effects, and hegemonic stability. Containment, first enunciated by Kennan, referred to the military and economic sequestration of the Soviet Union. Russia's historical geography not simply its cultural difference was invoked to give this argument scientific respectability. As Kennan (1947, 574) put it:

> The very teachings of Lenin himself require great caution and flexibility in the pursuit of Communist purposes. Again these precepts are fortified by the lessons of Russian history: of centuries of obscure battles between nomadic forces over the stretches of a vast unfortified plain. Here caution, circumspection, flexibility and deception are the valuable qualities, and their value finds natural appreciation in the Russian or oriental mind.

A policy of 'adroit and vigilant application of counter-force at a series of constantly shifting geographical and political points, corresponding to the shifts and maneuvers of Soviet policy' (ibid., 575) was necessary to contain the inherently expansive Soviet Union. Throughout his argument Kennan had recourse to a patriarchal mythology that repeatedly characterizes the Soviet Union as a potential seducer and rapist with repressed instincts that can burst forth at any point along its boundary unless there is constant pressure along all its length to keep it contained (O Tuathail and Agnew 1992).

Kennan's conception of containment was much more expansive than some later commentators have alleged. But it was at least confined to the margins of the Soviet Union, somewhat after the fashion of Mackinder's 'heartland model.' Although there is no evidence of a 'direct' link from Mackinder to Kennan, among some postwar American 'security intellectuals' Mackinder's ideas, designed for a very different historical context, took on a prophetic yet scientific role in naturalizing containment as a foreign and military policy.

However, the concept of containment became increasingly expansive. The 'geopolitical codes' of American Presidents underwent a series of shifts from the 1940s to the 1980s. Examining presidential State of the Union addresses for their global geographical content, O'Loughlin and Grant (1990, 527) report the following trend:

> In the 1940s and 1950s, emphasis was placed on the perceived threat to the 'Rimland,' that zone of containment arranged in a semicircle around the Soviet heartland. In the 1960s, attention to specific conflicts in Cuba and Vietnam was added to the dominant U.S./U.S.S.R. competitive theme. In the 1970s decade of detente, attention to foreign policy was reduced, only to be revived strongly in a regional guise in the late 1970s and early 1980s by Presidents Carter and Reagan. During the 1980s, the regional focus of U.S./U.S.S.R. competition has shifted to the Middle East, Southern Africa and Central America.

The main way in which the scope of containment was enlarged was through the logic of the so-called domino theory or domino effects. This argues that the sooner some potential threat to the global *status quo* was engaged, wherever it might occur, the less likely was it to produce a spread or contagion effect that could eventually lead all the way back to the United States. In a more sophisticated version the domino theory holds that the credibility of American commitments in more important regions, such as Europe, would be undermined by a failure to protect client regimes in distant corners of the globe. In such circumstances, American resolve to resist any aggression would be opened to doubt and the Great Adversary would be emboldened.

The metaphor of falling dominoes was first used by President Eisenhower in the mid-1950s to describe the consequences of the 'loss' of South Vietnam if the Communist insurgents there prevailed.

> You have a row of dominoes set up, you knock over the first one, and what will happen to the last one is the certainty that it will go over very quickly. So you could have the beginning of a disintegration that would have the most profound influences.
>
> (quoted in Gregory 1978, 275)

But President Truman had used much the same logic in arguing for American intervention in the Greek Civil War in the late 1940s. His metaphor was different:

> Like apples in a barrel infected by one rotten one, the corruption of Greece would infect Iran and all to the East. It would also carry infection to Africa through Asia Minor and Egypt, and to Italy and France, already threatened by the strongest domestic Communist

parties in Western Europe. The Soviet Union was playing one of the greatest gambles in history at minimal cost.

This logic was invoked again at the outset of the Korean War, repeatedly and explicitly during the Vietnam War, and in Chile and Angola in the 1970s. Finally, President Reagan used it as the center-piece of his administration's policy towards Central America in the 1980s. The successful overthrow of the Somoza dictatorship in Nicaragua in 1979 led the Reagan administration to support a group of counter-revolutionaries called the Contras committed to removing the Sandinista regime that had recently come to power and which was supported by the Soviet Union and its Caribbean ally, Cuba. This support was based primarily on the logic of the domino theory. President Reagan himself speculated openly that if the Sandinista government was not removed it would destabilize surrounding governments and very soon Sandinistas would show up at the Mexican border of the United States.

The domino theory is really a chain-reaction or contagion metaphor, whether in the rotten apple or falling dominoes version. It effectively externalizes local conflicts into aspects of the overarching global conflict. It does so by tying American national security into distant places through the possibility of the spread of revolution or the incubus of communism. Political change is thereby converted into a natural epidemiological process that threatens to diffuse like a disease into hitherto uninfected regions – eventually back to the United States itself. The fear of appeasement animating the generation that ran American (and Soviet) foreign policy after the Second World War was expressed in a linkage metaphor that put appeasement (or negotiation) beyond the pale of acceptable discourse.

As American dominance over its part of the globe slipped in the 1970s a third naturalizing prop was added to Cold War geopolitical discourse. This one involved arguing that inter-state cooperation and an optimal world economy needed a benevolent hegemon, such as the United States, for its best possible operation. 'Both Great Britain in the nineteenth century and the United States after the Second World War helped bring about an interdependent and overall peaceful world' (Grunberg 1990, 433). Whatever its empirical merit, and this is doubtful, this idea served to 'link the fate of the world with that of the United States,' there was 'no escape from chaos, except through a rejuvenation of U.S. power' (ibid., 447–8).

Rather like a kindly father or a self-sacrificing hero America's destiny was to hold off the cataclysm that would otherwise follow its demise. The justification for its centrality, however, was different. American images of fatherhood were just too ambiguous! Justification was offered in terms of the *systemic need* for a central provider of 'public goods' – lending at last resort, enforcing liberal trading rules, providing a stable unit of monetary account – that would be underprovided without the hegemon because it would be possible for any one state to consume its benefits without paying for them. In an anarchic international system only a benevolent despot can provide enforcement in the common interest. Consequently, the successful working of the world economy required continuing American 'leadership' (Plate 5.5).

The Cold War geopolitical imagination, therefore, although overtly ideological in the sense of competing commitments to different political–economic models also offered various naturalized 'fictions' that could be called up for periodic duty. They allowed the Cold War to become a self-fulfilling prophecy and placed it beyond rational consideration. Overall, the articulation of Cold War geopolitics helped to secure and reinforce a set of 'geographical identities ('the West,' 'the Soviet Union,' 'the United States'), while serving to discipline domestic social and cultural differences within these spaces' (O Tuathail 1993). The polarity between the United States and the Soviet Union was vital to this process. Even though exceptional intellectuals of statecraft, such as

Plate 5.5 Handing over to the world's policeman. A puzzled President Clinton adorned with the badge 'The World's Policeman' is handed the future safety of Hong Kong by a befeathered and grinning Tony Blair, after the return of the British colony to the jurisdiction of China
Source: Peter Brookes, The *Times*, July 1, 1997

President Nixon and his Secretary of State Henry Kissinger in the early 1970s, tried to move American foreign policy from its bipolar mode into a multipolar balance of power configuration, their emphasis on demonstrating 'credibility' wherever US power might be in question led them through the domino analogy right back to bipolarity.

AFTER IDEOLOGICAL GEOPOLITICS?

The demise of the Soviet Union as an ideological Other has undermined the entire basis of Cold War ideological geopolitics. Particularly in the United States the protective cocoon that the Cold War provided to political and intellectual life has shriveled. In its place has come an extreme ontological insecurity, a widespread sense of uncertainty about how to organize world politics in its absence. There is even a current of nostalgia for the 'good old days' when East was East and West was West and never the twain should meet. But whether ready substitutes will be found is doubtful. One point does seem clear, however: ideological geopolitics worked because two states pretending to the mantle of 'modernity' confronted one another globally. Neither a 'militant Islam' nor evil 'drug barons' provide an equally well-defined, competitive, and potent substitute, although some commentators are now offering demonized portraits of Japan and the Islamic world that would give this impression. The geopolitical imagination must once more be reconstituted.

The two most obvious candidates as the basis for a new geopolitics are one that identifies those new practices and representations of the deterritorializing and transnational global economy and another that sees the prospect of culture wars between different 'civilizations.' In the first case there is the possibility of using the plurality of spaces emerging under the influence of the transnational liberalism described in earlier chapters (particularly Chapters 3 and 4) for a kind of 'anti-geopolitics' built upon a commitment to treating places and people as if they counted independently of their global economic and military 'significance.' Absent of this subversion, there is likely to be a deepening of the 'market access' regime of world capitalism and its commitment to a world economy in which capital will be increasingly unconstrained in moving around the globe to exploit differences in rates of return on investment. A world of 'rich' and 'poor' zones, better accounted for in terms of the system of world cities and their hinterlands than by the world political map, will progressively replace the contemporary hierarchy of states (see Figure 5.2). Whichever trend becomes dominant, there is a limited long-term likelihood of returning to the contours of world politics as they were defined during the previous three ages of geopolitics. From this point of view, whatever happens will be something new in which the old principles will be of diminished relevance. In the interim, however, they will not readily

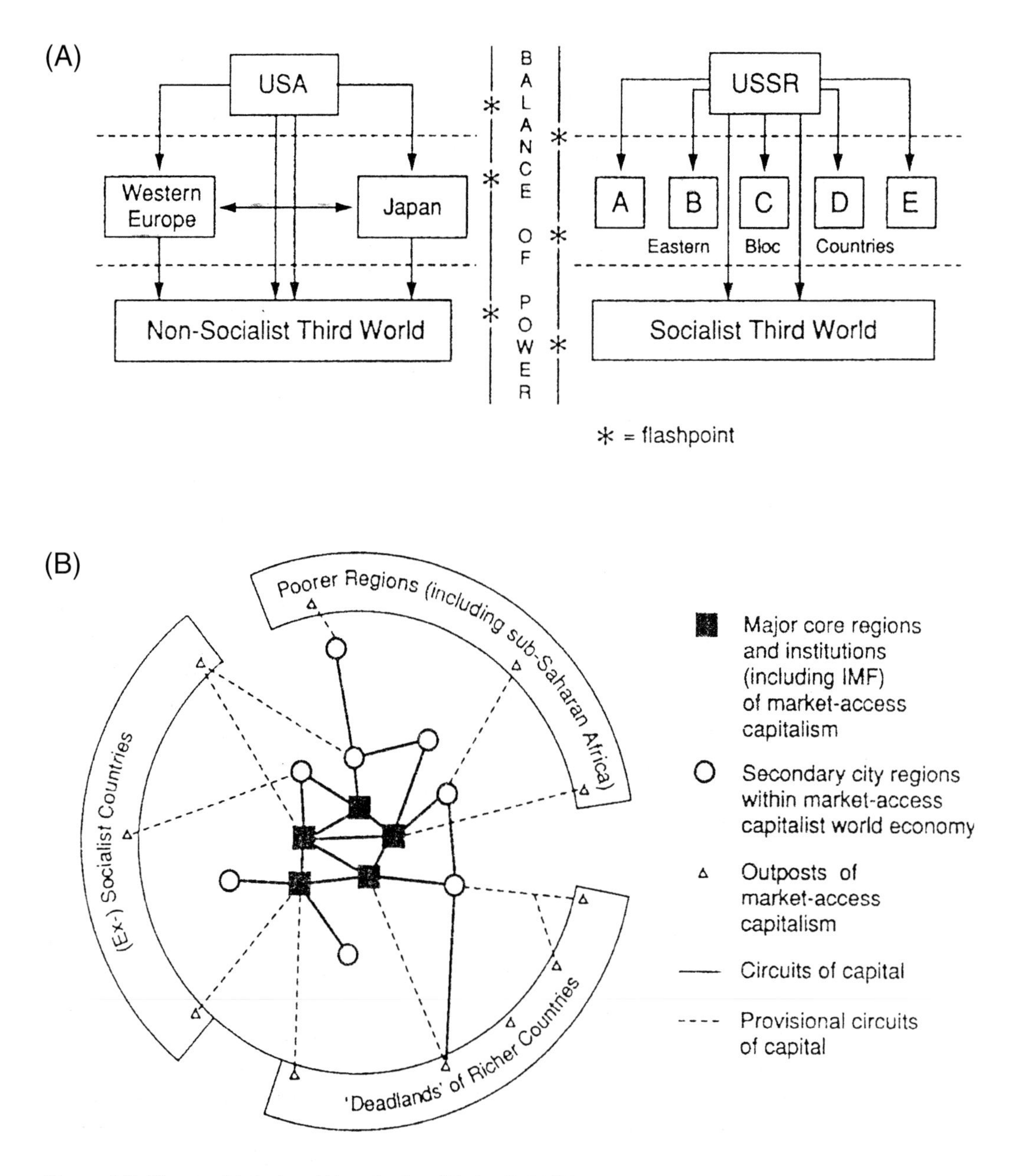

Figure 5.2 The world during (A) and after (B) the Cold War
Source: Agnew and Corbridge, 1995

disappear. Ideas about relative decline, state–territorial competition, plotting state strategy on a global basis, and thinking of development in terms of backwardness and modernity are so deep-seated as to defy easy supersession. The vocabulary of state-centered geopolitics is still underwritten materially by pressures from social groups reacting to increased economic globalization by reviving ethnic and local identities or attempting to resurrect state powers. Strong national identities die hard. States still provide the main 'opportunity structures' for most forms of political activity, even as 'contragovernmental' forces increase in number and scope. Governmentality (territorialized regimes of governmental authority) and movements/trends counter to this now exist in an unstable tension.

The 'culture wars' scenario has become particularly popular with those looking to reconstitute the ideological geopolitics of the Cold War on a multipolar basis. Cold War geopolitics rested on an essential opposition drawn between what were supposed to be two *completely different* types of society/culture. In reality, of course, cultural differences are always relative ones and there is much in common between 'cultures' that when isolated one from the other appear more particularistic than they really are. Both Islamic and Confucian-Asian cultures, for example, combine influences from other parts of the world (above all Europe) and have changed profoundly over the years. The image of fixed, isolated cultures should be rejected for the ideological imposition that it is. Yet, an increasing number of writers and intellectuals of statecraft are placing stress on the importance of cultural values and institutions in the confusion left in the wake of the Cold War. To the American political scientist Samuel Huntington (1993), for example, future wars will occur between the nations and groups of 'different civilizations' – western, Confucian, Japanese, Islamic, Hindu, Orthodox Christian, and Latin American, perhaps also African and Buddhist (see Figure 5.3). The 'fault-lines' between these cultures will define the geopolitical battle-lines of the future. 'Culture and cultural identities . . . are shaping the patterns of cohesion, disintegration and conflict in the post-cold war world . . . Global politics is being reconfigured along cultural lines' (Huntington 1993, 23).

The problem with this kind of scenario is twofold. First, who identifies with such broad-scale 'cultures'? Certainly, the 'western' category is problematic as contemporary attempts at creating a common sense of 'Europeanness' (in the European Union) founder on the shoals of revived and invented national and ethnic identities. Second, globalization undermines cultural 'closure.' The increasing flows of information, goods, people, and capital around certain parts of the world (particularly between Europe, North America, and East Asia) not only cause potential friction between cultures, they also tie cultures together, and increase tensions within culture areas as different social groups and individuals make different judgments about this or that external influence. Except on a superorganic view which sees them as external to populations, cultures are never set

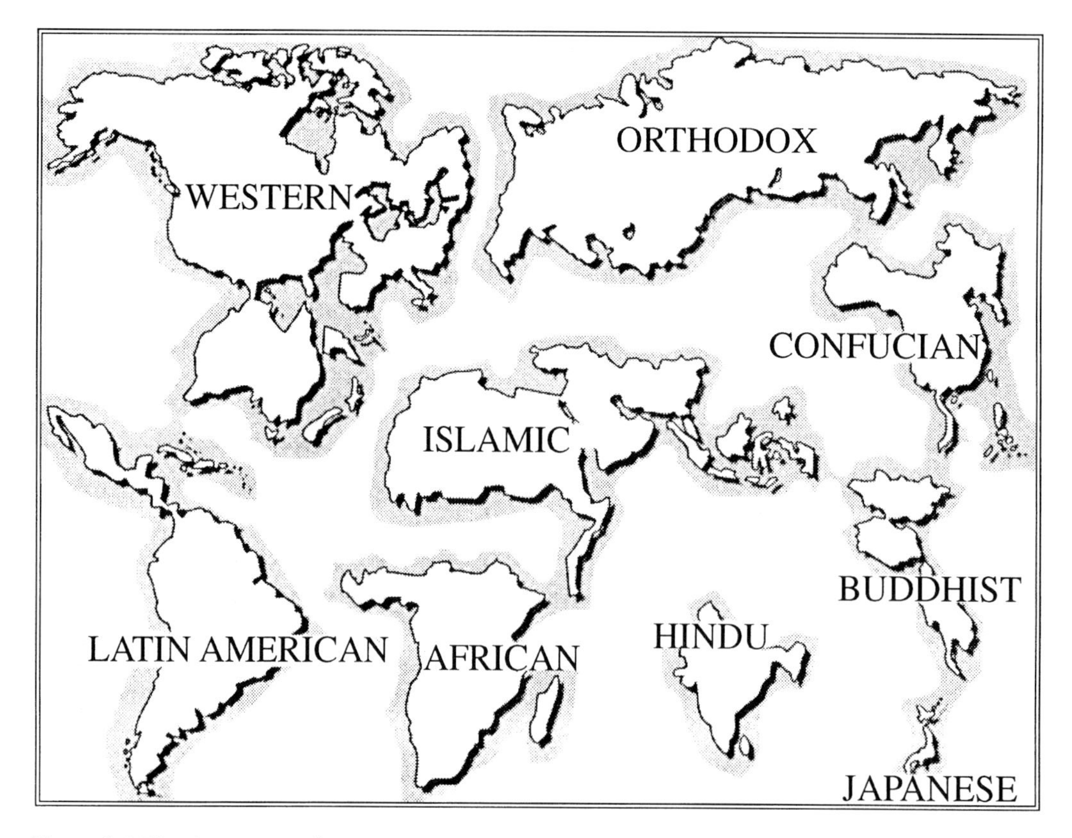

Figure 5.6 Huntington's civilizations (1993). Each civilization is identified as entirely discrete to visually convey the relative size and range of the civilizations as a set.

in stone; they adapt over time in response to external pressures and internal shifts. 'Culture' cannot, therefore, readily substitute for the all-embracing role of ideology in the Cold War. The increased scope for independent action by more states with the end of the Cold War and the pressures imposed by globalization work against it.

CONCLUSION

The modern geopolitical imagination, like the configuration of global space it purports to understand, was never set for all time. In this chapter I have surveyed how the modern geopolitical imagination has adapted to and underwritten the three eras of modern world politics. The civilizational geopolitics of the early nineteenth century gave way to the naturalized geopolitics of the late nineteenth and early twentieth centuries. After the Second World War an ideological geopolitics was constituted on the basis of the new geopolitical 'realities' of the time.

Each of these versions of the modern geopolitical imagination has had its own particular combination of elements. The first was based rather more on a sense of the opposition between European and other spaces that was then taken as justifying European predominance. The second focused particularly on the exclusive claims to territory and empire of competing states whose interests were economic and whose identities were racial. The third rested on the competing models of modernity offered by the United States and the Soviet Union, two states whose histories of expansion and internal organization differ profoundly from those of the classic European Great Powers.

But there have also been important continuities. These have been identified in the persisting themes of the geographical projection of 'backward vs. modern' flowing through geopolitical discourse from its origins in Renaissance Europe, seeing the world as a single entity and territorial states as the exclusive actors in world politics pursuing strategies of world primacy. The world can now be said to be living in postmodern times because of an increased awareness of the Eurocentric nature of this way of thinking. A key challenge is to survey and critically engage alternative scenarios for the future in as full knowledge as possible of the workings of the modern geopolitical imagination.

RECOMMENDED READING

Campbell, D. 1992 *Writing Security: United States Foreign Policy and the Politics of Identity*. Minneapolis: University of Minnesota Press.

Chow, R. 1996 We endure, therefore we are: survival, governance, and Zhang Yimou's *To Live*. *The South Atlantic Quarterly* 95: 1039–64.

Dalby, S. 1988 Geopolitical discourse: the Soviet Union as other. *Alternatives* 13: 415–42.

Horne, A. 1962 *The Price of Glory: Verdun 1916*. London: Penguin.

Huntington, S. P. 1993 The clash of civilizations? *Foreign Affairs* 72: 22–49.

Kern, S. 1983 *The Culture of Time and Space, 1880–1918*. Cambridge, MA: Harvard University Press.

Kristof, L. 1968 The Russian image of Russia, in C. A. Fisher (ed.) *Essays in Political Geography*. London: Methuen.

Pagden, A. 1994 *Lords of All the World: Ideologies of Empire in Spain, Britain and France c.1500–c.1800*. New Haven, CT: Yale University Press.

Smith, G. 1994 *The Last Years of the Monroe Doctrine*. New York: Hill and Wang.

Spurr, D. 1993 *The Rhetoric of Empire: Colonial Discourse in Journalism, Travel Writing, and Imperial Administration*. Durham, NC: Duke University Press.

Stephanson, A. 1995 *Manifest Destiny: American Expansion and the Empire of Right*. New York: Hill and Wang.

Taylor, P. J. (ed.) 1993 *The Political Geography of the Twentieth Century*. London: Belhaven Press.

6
CONCLUSION

•

From its origins in the late nineteenth century, political geography has been largely captive to a geopolitical imagination that has remained until recently unacknowledged and unexamined. Indeed, the founders of political geography, such as Ratzel, Kjellen and Mackinder, were major exponents of a particular variety of that imagination: the naturalized geopolitics identified in the previous chapter. The purpose of this book is to analytically trace the beginnings and development of the modern geopolitical imagination by exposing its component parts and showing how these came together under different material conditions and as a result of contestation to produce a succession of different geopolitical epochs.

Two theoretical premises inform the perspective of the book. The first is that in the realm of political geography the intellectual and the political are not separable. From its origins political geography has served statecraft, usually that of particular states. Yet, it has claimed to do so under the sign of objectivity; most typically in the form of the view from nowhere. This contradiction is more than a little problematic. I take the position that it undermines the entire approach. Rather like the explorers of Australia discussed in Chapter 1, political geographers have found what we wanted to find because we knew all along. What we knew kept the geopolitical treadmill turning indefinitely. We could never be surprised. We could never see that there was a native point of view that might be worth taking seriously. There was no possibility of searching for communalities and common understandings. We came, we knew, we conquered.

The second premise is that the making of the modern geopolitical imagination was not a once-only thing. Though the basic building-blocks or principles have remained more or less the same, the end product has gone through a number of important transformations as the political world it has purported to both reveal and reflect has changed. An imagination or ideas do not exist 'out there' or simply in texts and documents, they are implicit in practices or social action. To survive and prosper they must be passed on from generation to generation as a form of common sense or guide to action and must adapt successfully to challenges and changed historical contexts.

These premises and the arguments I develop from them are somewhat different from other recent commentaries on political geography, world politics and geopolitics. Of first importance, a modern geopolitical imagination is identified as preceding the first use of the word geopolitics itself. I see no problem with this. Only an extreme linguistic nominalism that endows words with a singular causal power would have it otherwise. I present a historical argument to the effect that geographical understandings of a certain type have informed the actions of political élites and that political geography (and other fields) have taken these understandings as the basis for their understanding. Although textual analysis of writers explicitly using the term geopolitics is certainly not excluded by this approach, remaining solely concerned with it strikes me as limiting. The role of the modern geopolitical imagination in world politics is best shown by its practical consequences over a long period of time rather than by examining the claims of its disciplinary exponents. A challenge for some contemporary writers is to decide whether they are primarily interested in the history of geographic thought or in the history of the modern geopolitical imagination, or at least to be clear about which it is.

The approach that I take to understanding the making of the modern geopolitical imagination is a historicist one; it does not presume a set of fixed beliefs, knowledge, attitudes and practices that remain essentially unchanged down the centuries. To the contrary, it presumes change in the content of the geopolitical imagination as a result of changing times. Changing technological, economic, social and political conditions are particularly important. This too is a controversial position. The tendency is to cite some feature from the distant or recent past, say, understandings of the meanings of racial categories, and then project these forward or backward in time as if they were structural features of thought rather than ones with distinctive historical trajectories. In each chapter and, above all, in Chapter 5, I have tried to indicate how dependent on historical context the content and deployment of the modern geopolitical imagination have been. Whatever the deficiencies of periodization used in Chapter 5, the appending of the term 'modern' to 'geopolitical imagination' suggests the strong sense of historical dependence that I believe any accounting of the geographical understandings at work in world politics must presume. This is readily accepted when considering the recent breakdown of Cold War geopolitics, as signaled by such phrases as 'rethinking geopolitics,' but it seems harder to accept when examining the past. The fear of criticism of whatever periodization one adopts seems to provoke hostility to thinking historically at all.

A theme running through all of the chapters is that present conditions are not propitious for the continuation of the geopolitical imagination as we have known it in the past though there are continuing attempts at breathing new life into it (see, for example, Brzezinksi 1997). Indeed, the fact that it has lost much of its intellectual magic, if not yet all of its political appeal, suggests that the possibility of rethinking geopolitics also depends on changed historical conditions. It is certainly an appropriate time to rethink political geography's historical dependence on the geopolitical imagination. Again, other commentators would differ. In my view, contemporary conditions suggest the need to abandon the *a priori* commitment to the global and national-state geographical scales as somehow exhausting the sum total of causal influences on world politics. This means rejecting both the *status quo* and the search for a 'new' geopolitical imagination that simply builds on the old foundations. Rather, what is needed is a geographical imagination that takes places seriously as the settings for human life and tries to understand world politics in terms of its impacts on the material welfare and identities of people in different places. This involves addressing questions of national and other identities under conditions of massive population movement and diaspora; growing global inequalities; the increased role of regions and localities in regulating economic growth; the rise of supranationalism on world-regional and international scales; and, above all, the growing impacts of globalization of production and finance on states and localities around the world. This research agenda will be subverted, however, if it is addressed using the elements of the modern geopolitical imagination. After two generations of state-centricity and global geographical determinism, and approaching the centennial of the first use of the word 'geopolitics,' political geographers must finally choose whether to be agents of an imagination that has imposed manifold disasters on humanity or to try to understand geographical communalities and differences in their own right. In other words, it is time to choose sides.

RECOMMENDED READING

Agnew, J. 1993 Representing space: space, scale and culture in social science, in J. Duncan and D. Ley (eds) *Place/Culture/Representation*. London: Routledge.

Agnew, J. 1995 Democracy and human rights after the Cold War, in R. J. Johnston, P. J. Taylor and M. J. Watts (eds) *Geographies of Global Change*. Oxford: Blackwell.

Brzezinski, Z. 1997 *The Grand Chessboard: American Primacy and its Geostrategic Imperatives*. New York: Basic Books.

Entrikin, J. N. 1991 *The Betweenness of Place: Towards a Geography of Modernity*. Baltimore: Johns Hopkins University Press.

Harris, N. 1995 *The New Untouchables: Immigration and the New World Worker*. London: Penguin.

Harvey, D. 1989 *The Condition of Postmodernity*. Oxford: Blackwell.

Harvie, C. 1994 *The Rise of Regional Europe*. London: Routledge.

Knox, P. and Agnew, J. 1998 *The Geography of the World Economy*. 3rd edition, London: Arnold.

O Tuathail, G. 1996 *Critical Geopolitics: The Politics of Writing Global Space*. Minneapolis: University of Minnesota Press.

Scott, A. J. 1996 Regional motors of the world economy. *Futures*, 28: 391–411.

Williams, R. 1986 *Towards the Year 2000*. New York: Pantheon.

GLOSSARY OF TERMS

•

cosmography The science of the world in which the structure of the cosmos in general and the world in particular was to be accounted for in an exhaustive inventory of particular features, including folklore as well as documented observation, but always in terms of a pre-existing imagined world map which was to be filled in.

discourse The setting or domain in which words are used and take on specific meanings. In geopolitical discourse words such as 'backward,' 'race,' 'heartland,' and 'Great Power' take on meaning in relation to other words within a stream of understandings that provides guidance to and interpretation of practices or action.

élites The communities of government officials, political leaders, foreign-policy experts and their immediate supporters throughout the world who conduct, influence and comment upon world politics.

geopolitical imagination, modern The view of the world and its geographical workings that accompanied the rise of the state and capitalism in Europe and that was both stimulated by and informed the European encounter with the rest of the world.

geopolitics The study of the impact of geographical distributions and divisions on the conduct of world politics. In its original usage it referred to the impact on inter-state relations of the spatial disposition of continents and oceans and the distribution of natural and human resources. Today, however, the term also covers examination of all of the geographical assumptions, designations and understandings that enter into the making of world politics (as in critical geopolitics).

Great Powers The dominant or leading states and empires of a particular epoch.

hegemony The dominant understandings and rules governing political and economic practice of a given period as accepted by coalitions of élites. Sometimes associated in world politics with the dominance exerted by a single state (or hegemon). This usage also accepts that dominance is rarely simply coercive but extends into control over the 'rules of the game' or agenda setting.

ideology A structured understanding of how the world works and in whose interests. In one usage ideology is reserved for one's political opponents, implying that it is false or misleading thought. Another sees any pattern of thought directed towards 'unmasking' the world as it appears as an ideology.

mercantilism An economic doctrine prevalent in Europe in the sixteenth to eighteenth centuries in which states strongly regulated industry and trade on the basis of the following four beliefs: that exports to foreign countries increase national strength; that exports are preferable to internal trade and foreign imports; that state wealth depends primarily upon the possession of gold and silver; and that state intervention is justified to achieve these objectives. Mercantilist ideas never entirely disappeared. They crop up in connection with most territorial modes of economic organization, including the exploitation of colonies to provide raw materials and precious metals.

nation A group of people of common nationality (kinship or civic bonds being the most common basis to the definition of nationality) occupying and claiming a historic territory.

nation–state A state in which the territorial boundaries also enclose a claimed nation which the state represents.

polity A structured organization of power within a social group that has a distinctive identity, can mobilize group members and persists over time. By this definition not only territorially bounded states but churches, interest groups and social movements can be types of polity. Polities need not be organized territorially and can co-exist in layered, overlapping and interacting webs of power and influence.

power The capacity to achieve goals. Often, the power to coerce others (or power *over*) is privileged definitionally at the expense of the power of agency (or power *to*).

spatiality The way in which terrestrial space is thought of as affecting the organization of a given phenomenon such as power or social relations.

state The government of a bounded territory which claims common political and legal authority and a monopoly of legitimate force and other sovereign powers throughout its jurisdiction. This is often fused with the idea of nation (a common people) to create the nation–state, in which there is a territorial matching of nation and state.

state territoriality The association of the state with a discrete and well-defined territory.

Three Worlds The three worlds of development as defined in the Cold War: the First World of modern, capitalist states, the Second World of modern but communist states, and the Third World in which the first two vied for influence.

transnational liberalism A set of ideas and practices associated with economic organization that is not confined by state boundaries but which incorporates the whole world as a potential space for investment, trade and production.

Westphalian system The set of mutually recognized states that slowly emerged as a result of the Treaty of Westphalia (1648) in which the right of a ruler to an exclusive territorial jurisdiction was first legislated. As new states (including old empires and other polities) have appeared over the years they have wanted to be recognized as parts of this system, even though the European original relied on a close approximation between nation and state. The increased permeability of state boundaries as the world economy globalizes, the lack of effectiveness and legitimacy of many states and the lack of matching between nations and states, all undermine the primary requirement of exclusive jurisdiction.

BIBLIOGRAPHY

•

INTRODUCTION

Agnew, J. A. 1994 The territorial trap: the geographical assumptions of international relations theory. *Review of International Political Economy*, 1: 53–80.

Mackinder, H. J. 1904 The geographical pivot of history. *Geographical Journal*, 13: 421–37.

McNamara, R. J. 1995 *In Retrospect: The Tragedy and Lessons of Vietnam*. New York: Times Books.

O Tuathail, G. 1996 *Critical Geopolitics: The Politics of Writing Global Space*. Minneapolis: University of Minnesota Press.

Wallerstein, I. 1991 *Unthinking Social Science: The Limits of Nineteenth-Century Paradigms*. Cambridge: Polity Press.

CHAPTER 1

Berger, J. 1972 *Ways of Seeing*. London: BBC Books.

Carpenter, R. H. 1995 *History as Rhetoric: Style, Narrative and Persuasion*. Columbia, SC: University of South Carolina Press.

Cosgrove, D. E. 1996 Geography and vision. Inaugural lecture, Department of Geography, Royal Holloway, University of London, February 29.

Crary, J. 1990 *Techniques of the Observer: On Vision and Modernity in the Nineteenth Century*. Cambridge, MA: MIT Press.

Crocker, C. 1977 The social function of rhetorical forms, in J. D. Sapir and C. Crocker (eds) *The Social Uses of Metaphor: Essays on the Anthropology of Rhetoric*. Philadelphia: University of Pennsylvania Press.

Crone, G. R. 1978 *Maps and Their Makers: An Introduction to the History of Cartography*. Hamden, CT: Archon Books.

Crosby, A. W. 1997 *The Measure of Reality: Quantification and Western Society, 1250–1600*. Cambridge: Cambridge University Press.

Dijkink, G. 1996 *National Identity and Geopolitical Visions: Maps of Pride and Pain.* London: Routledge.

Elliott, J. H. 1991 The world after Columbus. *New York Review of Books*, 10 October: 10–14.

Gopnik, A. 1996 The first Frenchman. *The New Yorker*, 7 October: 44–53.

Greenblatt, S. 1991 *Marvelous Possessions: The Wonder of the New World.* Chicago: University of Chicago Press.

Gregory, D. 1994 *Geographical Imaginations.* Oxford: Blackwell.

Hannaford, I. 1996 *Race: The History of an Idea in the West.* Baltimore, MD: Johns Hopkins University Press.

Harley, J. B. 1989 Deconstructing the map. *Cartographica*, 26: 1–20.

Huntington, S. P. 1993 The clash of civilizations? *Foreign Affairs*, 72: 22–49.

Ivie, R. L. 1984 Speaking 'common sense' about the Soviet threat: Reagan's rhetorical stance. *Western Journal of Speech Communication*, 48: 39–50.

Jay, M. 1993 *Downcast Eyes: The Denigration of Vision in Twentieth Century French Thought.* Berkeley and Los Angeles: University of California Press.

Kemp, M. 1990 *The Science of Art: Optical Themes in Western Art from Brunelleschi to Seurat.* New Haven: Yale University Press.

Kern, S. 1983 *The Culture of Time and Space, 1880–1918.* Cambridge, MA: Harvard University Press.

Knox, P. L. and Agnew, J. A. 1994 *The Geography of the World Economy.* 2nd Edition, London: Edward Arnold.

Kupperman, K. O. 1995 Introduction: the changing definition of America, in K. O. Kupperman (ed.) *America in European Consciousness, 1493–1750.* Chapel Hill, NC: University of North Carolina Press.

Lestringant, F. 1994 *Mapping the Reniassance World: The Geographical Imagination in the Age of Discovery.* Berkeley and Los Angeles: University of California Press.

Manchester, W. 1992 *A World Lit Only by Fire. The Medieval Mind and the Renaissance, Portrait of an Age.* Boston: Little, Brown.

Mattelart, A. 1996 *The Invention of Communication.* Minneapolis: University of Minnesota Press.

Ninkovich, F. 1994 *Modernity and Power: A History of the Domino Theory in the Twentieth Century.* Chicago: University of Chicago Press.

Nussbaum, F. 1995 *Torrid Zones: Maternity, Sexuality, and Empire in Eighteenth-Century English Narratives.* Baltimore, MD: Johns Hopkins University Press.

Oakes, G. 1995 *The Imaginary War: Civil Defense and American Cold War Culture.* New York: Oxford University Press.

Pagden, A. 1994 *Lords of All the World: Ideologies of Empire in Spain, Britain and France c.1500–c.1800*. New Haven: Yale University Press.

Pletsch, C. E. 1981 The Three Worlds, or the division of social scientific labor, circa 1950–1975. *Comparative Studies in Society and History*, 23: 565–90.

Ryan, M. T. 1981 Assimilating New Worlds in the sixteenth and seventeenth centuries. *Comparative Studies in Society and History*, 23: 519–38.

Said, E. W. 1978 *Orientalism: Western Conceptions of the Orient*. New York: Vintage.

Simpson, D. 1993 *Romanticism, Nationalism and the Revolt Against Theory*. Chicago: University of Chicago Press.

Simpson, D. 1996 *The Academic Postmodern and the Rule of Literature: A Report on Half-Knowledge*. Chicago: University of Chicago Press.

Solomon, R. C. 1993 *The Bully Culture: Enlightenment, Romanticism, and the Transcendental Pretense*. Lanham, MD: Littlefield Adams.

Springborg, P. 1992 *Western Republicanism and the Oriental Prince*. Cambridge: Polity Press.

Virilio, P. 1994 *The Vision Machine*. Bloomington, IN: Indiana University Press.

Wolf, E. R. 1982 *Europe and the People Without History*. Berkeley and Los Angeles: University of California Press.

Wolff, L. 1994 *Inventing Eastern Europe: The Map of Civilization in the Mind of the Enlightenment*. Stanford, CA: Stanford University Press.

Wolter, J. A. and Grim, R. E. (eds) 1996 *Images of the World: The Atlas Through History*. New York: McGraw-Hill/Library of Congress.

CHAPTER 2

Appadurai, A. 1988 Putting hierarchy in its place. *Cultural Anthropology*, 3: 36–49.

Applegate, C. 1990 *A Nation of Provincials: The German Idea of* Heimat. Berkeley and Los Angeles: University of California Press.

Banfield, E. 1958 *The Moral Basis of a Backward Society*. New York: Free Press.

Bondanella, P. and Musa, M. (eds) 1979 *The Portable Machiavelli*. New York: Penguin.

Boroujerdi, M. 1996 *Iranian Intellectuals and the West: The Tormented Triumph of Nativism*. Syracuse, NY: Syracuse University Press.

Carini, C. 1975 *Benedetto Croce e il partito politico*. Florence: Olschki.

Carter, P. 1989 *The Road to Botany Bay: An Essay on Landscape and History*. Chicago: University of Chicago Press.

Chow, R. 1996 We endure, therefore we are: survival, governance, and Zhang Yimou's *To Live*. *The South Atlantic Quarterly*, 95: 1039–64.

Duncan, J. 1993 Sites of representation: place, time and the discourse of the other, in J. Duncan and D. Ley (eds) *Place/Culture/Representation*. London: Routledge.

Eagleton, T. 1991 *Ideology: An Introduction*. London: Verso.

Estevo, G. 1992 Development, in W. Sachs (ed.) *The Development Dictionary: A Guide to Knowledge as Power*. London: Zed Books.

Eze, E. C. (ed.) 1997 *Race and the Enlightenment: A Reader*. Oxford: Blackwell.

Fabian, J. 1983 *Time and the Other: How Anthropology Makes its Object*. New York: Columbia University Press.

Ferrera, M. 1987 Il mercato politico-assistenziale, in U. Ascoli and R. Catanzaro (eds) *La società italiana degli anni ottanta*. Bari: Laterza.

Forgacs, D. 1990 *Italian Culture in the Industrial Era, 1880–1980: Cultural Industries, Politics and the Public*. Manchester: Manchester University Press.

Ginsborg, P. 1990 *A History of Contemporary Italy: Society and Politics, 1943–1988*. London: Penguin.

Goody, J. 1996 *The East in the West*. Cambridge: Cambridge University Press.

Gramsci, A. 1971 *Selections from the Prison Notebooks*. London: Lawrence and Wishart.

Guha, R. 1989 Dominance without hegemony and its historiography. *Subaltern Studies*, 6: 210–309.

Gupta, A. and Ferguson, J. 1992 Beyond 'culture': space, identity, and the politics of difference. *Cultural Anthropology*, 7: 6–23.

Hegel, G. W. F. 1821/1967 *Philosophy of Right*. Oxford: Oxford University Press.

Helms, M. W. 1988 *Ulysses' Sail: An Ethnographic Odyssey of Power, Knowledge, and Geographical Distance*. Princeton, NJ: Princeton University Press.

Jacobitti, E. E. 1981 *Revolutionary Humanism and Historicism in Modern Italy*. New Haven: Yale University Press.

Jacobs, J. M. 1996 *Edge of Empire: Postcolonialism and the City*. London: Routledge.

Kermode, F. 1967 *The Sense of an Ending*. New York: Oxford University Press.

Lanaro, S. 1989 *L'Italia nuova. Identità e sviluppo, 1861–1988*. Turin: Einaudi.

Lembo, R. 1988 Il Mezzogiorno tra storia e antropologia. *Studi Storici*, 29: 1051–68.

Lugard, F. D. 1926 *The Dual Mandate in Tropical Africa*. Edinburgh: Oliver and Boyd.

Mack Smith, D. 1994 *Mazzini*. New Haven: Yale University Press.

Maier, C. J. 1979 *La rifondazione dell'Europa borghese: Francia, Germania e Italia nei decennio successivo alla prima guerra mondiale*. Bari: De Donato.

Mandrou, R. 1978 *From Humanism to Science, 1480–1700*. London: Penguin.

Mantino, A. 1953 *La formazione della filosofia politica di Benedetto Croce*. Bari: Laterza.

Manzoni, A. 1827/1972 *I promessi sposi/ The Betrothed*, trans. Bruce Penman. London: Penguin.

Mason, T. 1988 Italy and modernization: a montage. *History Workshop*, 25/26: 127–47.

Mitchell, T. L. 1839 *Three Expeditions into the Interior of Eastern Australia: with descriptions of the Recently Explored Regions of Australia Felix and of the Present Colony of New South Wales*. 2 vols, 2nd edition, London: T. W. Boone.

Pagden, A. 1992 *European Encounters with the New World: From Renaissance to Romanticism*. New Haven: Yale University Press.

Parsons, T. 1951 *The Social System*. New York: Free Press.

Pletsch, C. E. 1981 The Three Worlds, or the division of social scientific labor, circa 1950–1975. *Comparative Studies in Society and History*, 23: 565–90.

Pratt, M. L. 1992 *Imperial Eyes: Travel Writing and Transculturation*. London: Routledge.

Putnam, R. 1993 *Making Democracy Work: Civic Traditions in Modern Italy*. Princeton, NJ: Princeton University Press.

Ryan, M. T. 1981 Assimilating New Worlds in the sixteenth and seventeenth centuries. *Comparative Studies in Society and History*, 23: 519–38.

Ryan, S. 1996 *The Cartographic Eye: How Explorers Saw Australia*. Cambridge: Cambridge University Press.

Said, E. W. 1978 *Orientalism: Western Conceptions of the Orient*. New York: Vintage.

Sartori, G. 1966 *Stato e politica nel pensiero di Benedetto Croce*. Naples: Morano.

Schwoebel, R. 1967 *The Shadow of the Crescent: The Renaissance Image of the Turk (1453–1517)*. Nieukoop: De Graaf.

Scott, J. T. and Sullivan, V. B. 1994 Patricide and the plot of *The Prince*: Cesare Borgia and Machiavelli's Italy. *American Political Science Review*, 88: 887–900.

Shapiro, M. J. 1995 The ethics of encounter: unreading/unmapping the Imperium. Paper presented at the International Studies Association Annual Meeting, Chicago, February.

Signorelli, A. 1986 Review of *La nostra Italia*. *L'Indice*, 8: 46.

Soja, E. 1989 *Postmodern Geographies: The Reassertion of Space in Social Theory*. London: Verso.

Springborg, P. 1992 *Western Republicanism and the Oriental Prince*. Cambridge: Polity Press.

Tipps, D. 1973 Modernization theory and the comparative study of societies: a critical perspective. *Comparative Studies in Society and History*, 15: 199–226.

Tullio-Altan, C. 1986 *La nostra Italia. Arretratezza socioculturale, clientelismo, trasformismo e ribellismo dall' Unità ad oggi*. Milan: Feltrinelli.

Veyne, P. 1990 *Bread and Circuses: Historical Sociology and Political Pluralism*. London: Allen Lane.

Weber, M. 1978 *Economy and Society*, Vol. 1, eds G. Roth and C. Wittich. Berkeley and Los Angeles: University of California Press.

Wolin, S. 1985 Postmodern society and the absence of myth. *Social Research*, 52: 217–39.

CHAPTER 3

Agnew, J. A. 1989 The devaluation of place in social science, in J. Agnew and J. Duncan (eds) *The Power of Place: Bringing Together Geographical and Sociological Imaginations*. Boston: Unwin Hyman.

Agnew, J. A. 1994 The territorial trap: the geographical assumptions of international relations theory. *Review of International Political Economy*, 1: 53–80.

Allen, B. 1995 From multiplicity to multitude: universal systems of deformation. *Symposium*, 49: 93–113.

Barry, K. 1997 Paper money and English Romanticism. Literary side-effects of the last invasion of Britain. *Times Literary Supplement*, 21 February: 14–16.

Billig, M. 1995 *Banal Nationalism*. London: Sage.

Bourdieu, P. 1977 *Outline of a Theory of Practice*. Cambridge: Cambridge University Press.

Brantlinger, P. 1996 *Fictions of State: Culture and Credit in Britain, 1694–1994*. Ithaca, NY: Cornell University Press.

Burch, K. 1994 The 'properties' of the state system and global capitalism, in S. Rosow, N. Inayatullah and M. Rupert (eds) *The Global Economy as Political Space*. Boulder, CO: Lynne Rienner.

Castells, M. 1996 *The Rise of the Network Society*. Oxford: Blackwell.

Cerny, P. G. 1993 The deregulation and re-regulation of financial markets in a more open world, in P. G. Cerny (ed.) *Finance and World Politics: Markets, Regimes and States in the Post-hegemonic Era*. Aldershot: Elgar.

Cohen, B. 1977 *Organizing the World's Money*. New York: Basic Books.

Cowhey, P. F. and Aronson, J. D. 1993 *Managing the World Economy: The Consequences of Corporate Alliances*. New York: Council on Foreign Relations Press.

Cox, R. W. 1987 *Power, Production, and World Order: Social Forces in the Making of History*. New York: Columbia University Press.

Deudney, D. 1995 Nuclear weapons and the waning of the *real*-state. *Daedalus*, 124: 209–31.

Deudney, D. 1996 Binding sovereigns: authorities, structures, and the geopolitics of the Philadelphian system, in T. Biersteker and C. Weber (eds) *State Sovereignty as Social Construct*. Cambridge: Cambridge University Press.

Dodd, N. 1995 Money and the nation–state: contested boundaries of monetary sovereignty in geopolitics. *International Sociology*, 10: 139–54.

Giddens, A. 1979 *Central Problems in Social Theory: Action, Structure and Contradiction in Social Analysis*. London: Macmillan.

Gray, A. 1993: *Ten Tales Tall and True*. San Diego: Harcourt Brace Jovanovich.

Helleiner, E. 1994 *States and the Reemergence of Global Finance: From Bretton Woods to the 1990s*. Ithaca, NY: Cornell University Press.

Helleiner, E. 1996 Historicizing territorial currencies: monetary structures, sovereignty and the nation–state. Paper presented at the International Studies Association, Annual Meeting, San Diego, 17–20 April.

Hudson, A. C. 1996 Globalization, regulation and geography: the development of the Bahamas and the Cayman Islands as offshore financial centres. unpublished PhD dissertation, University of Cambridge.

Julius, D. 1990 *Global Companies and Public Policy: The Growing Challenge of Foreign Direct Investment*. New York: Council on Foreign Relations.

Keohane, R. O. 1984 *After Hegemony: Cooperation and Discord in the World Political Economy*. Princeton, NJ: Princeton University Press.

Krishna, S. 1994 Cartographic anxiety: mapping the body politic in India. *Alternatives*, 19: 507–21.

Kundera, M. 1984 *The Unbearable Lightness of Being*. New York: Harper and Row.

Lukács, G. 1971 *The Theory of the Novel*. Cambridge, MA: MIT Press.

Lukes, S. 1975 *Power: A Radical View*. London: Macmillan.

Mann, M. 1984 The autonomous power of the state: its origins, mechanisms and results. *European Journal of Sociology*, 25: 185–213.

Massey, D. 1993 Politics and space/time. *New Left Review*, April: 65–84.

Melvin, M. 1988 The dollarization of Latin America as a market-enforced monetary reform: evidence and implications. *Economic Development and Cultural Change*, 36: 543–58.

Mosse, G. 1975 *The Nationalization of the Masses*. New York: Howard Fertig.

O'Brien, R. 1992 *Global Financial Integration: The End of Geography*. New York: Council on Foreign Relations Press.

Piron, S. 1996 Monnaie et majesté royale dans la France du 14e siècle. *Annales (Histoires, Sciences Sociales)*, 51, 2: 325–54.

Rosecrance, R. 1996 The virtual state. *Foreign Affairs*, 75: 584–604.

Rostow, W. 1960 *The Stages of Economic Growth: A Non-Communist Manifesto*. New York: Cambridge University Press.

Rotstein, A. and Duncan, C. 1991 For a second economy, in D. Drache and M. Gertler (eds) *The New Era of Global Competition*. Montreal: McGill-Queen's University Press.

Ruggie, J. G. 1993 Territoriality and beyond: problematizing modernity in international relations. *International Organization*, 47: 139–74.

Sandholtz, W. 1993 Choosing union: monetary politics and Maastricht. *International Organization*, 47: 1–39.

Spruyt, H. 1994: *The Sovereign State and its Competitors: An Analysis of Systems Change*. Princeton, NJ: Princeton University Press.

Thomson, J. E. 1994 *Mercenaries, Pirates, and Sovereigns: State-Building and Extraterritorial Violence in Early Modern Europe*. Princeton, NJ: Princeton University Press.

Trigilia, C. 1991 The paradox of the region: economic regulation and the representation of interests. *Economy and Society*, 20: 306–27.

Walker, R. B. J. 1993 *Inside/Outside: International Relations as Political Theory*. Cambridge: Cambridge University Press.

Waltz, K. E. 1979 *Theory of International Politics*. New York: Random House.

Watt, I. 1957 *The Rise of the Novel*. Berkeley and Los Angeles: University of California Press.

CHAPTER 4

Agnew, J. A. 1993 The United States and American hegemony, in P. J. Taylor (ed.) *The Political Geography of the Twentieth Century*. London: Belhaven Press.

Avishai, B. 1995 The inseparables: for Israel, separation from the Palestinians is a chimera. *The New Yorker*, 16 October: 5–6.

Brzezinski, Z. 1986 *Game Plan*. Boston: Atlantic Books.

Buzan, B. and Little, R. 1996 Reconceptualizing anarchy: structural realism meets world history. *European Journal of International Relations*, 2: 403–38.

Cox, R. W. 1992 Multilateralism and world order. *Review of International Studies*, 18: 161–80.

Department of Defense (DoD) 1995 *Annual Report*. Washington, DC: US Government Printing Office.

Deudney, D. 1996 Binding sovereigns: authorities, structures, and the geopolitics of the Philadelphian system, in T. Biersteker and C. Weber (eds) *State Sovereignty as Social Construct*. Cambridge: Cambridge University Press.

Frank, R. H. and Cook, P. J. 1995 *The Winner-Take-All Society*. New York: Free Press.

Gilpin, R. 1981 *War and Change in World Politics*. Cambridge: Cambridge University Press.

Hegel, G. W. F. 1931 *The Phenomenology of Mind*. 2nd English edition, New York: Humanities Press.

Inayatullah, N. 1997 Theories of spontaneous disorder. *Review of International Political Economy*, 4: 319–48.

Johnson, P. 1976 *A History of Christianity*. London: Weidenfeld and Nicolson.

Julius, D. 1990 *Global Companies and Public Policy: The Growing Challenge of Foreign Direct Investment*. New York: Council on Foreign Relations.

Kahler, M. 1992 Multilateralism with small and large numbers. *International Organization*, 46: 681–708.

Kennedy, P. 1987 *The Rise and Fall of the Great Powers: Economic Change and Military Conflict from 1500 to 2000*. New York: Random House.

Kennedy, P. 1995 'Too serious a business,' a reply to Professor Krugman. *Peace Economics, Peace Science and Public Policy*, 2: 16–22.

Klotz, A. 1995 Norms reconstituting interests: global racial equality and US sanctions against South Africa. *International Organization*, 49: 451–78.

Krueger, A. 1992 Global trading prospects for the developing countries. *The World Economy*, 15: 457–74.

Krugman, P. 1995 A reply. *Peace Economics, Peace Science and Public Policy*, 2: 26–30.

O Tuathail, G. 1993 Japan as threat: geo-economic discourses on the US–Japan relationship in US civil society, 1987–1991, in C. H. Williams (ed.) *The Political Geography of the New World Order*. London: Belhaven Press.

Porter, M. 1990 *The Competitive Advantage of Nations*. London: Macmillan.

Reich, R. 1991 The myth of "Made in the U.S.A." *Wall Street Journal*, 5 July: A6.

Ringmaar, E. 1996 On the ontological status of the state. *European Journal of International Relations*, 2: 439–66.

Rosenau, J. N. 1990 *Turbulence in World Politics: A Theory of Change and Continuity*. Princeton, NJ: Princeton University Press.

Scott, A. J. 1996 Regional motors of the global economy. *Futures*, 28: 391–411.

Shapiro, M. J. 1989 Representing world politics: the sport/war intertext, in J. Der Derian and M. J. Shapiro (eds) *International/Intertextual Relations*. Lexington, MA: Lexington Books.

Van Creveld, M. 1993 *On Future War*. Oxford: Blackwell.

Weldes, J. 1996 Constructing national interests. *European Journal of International Relations*, 2: 275–318.

Zacher, M. 1992 The decaying pillars of the Westphalian temple: implications for international order and governance, in J. N. Rosenau and E.-N. Czempiel (eds) *Governance Without Government: Order and Change in World Politics*. Cambridge: Cambridge University Press.

CHAPTER 5

Achebe, C. 1975 *Morning Yet on Creation Day*. London: Faber.

Agnew, J. A. and Corbridge, S. 1995 *Mastering Space: Hegemony, Territory and International Political Economy*. London: Routledge.

Applegate, C. 1990 *A Nation of Provincials: The German Idea of* Heimat. Berkeley and Los Angeles: University of California Press.

Barnouw, D. 1990 *Visible Spaces: Hannah Arendt and the German-Jewish Experience*. Baltimore, MD: Johns Hopkins University Press.

Bassin, M. 1987 Race contra space: the conflict between German *Geopolitik* and National Socialism. *Political Geography Quarterly*, 6: 115–34.

Bassin, M. 1991 Russia between Europe and Asia: the ideological construction of geographical space. *Slavic Review*, 50: 1–17.

Bassin, M. 1993 Turner, Solov'ev and the "frontier hypothesis": the nationalist signification of open spaces. *Journal of Modern History*, 65: 473–511.

Benn, S. I. 1967 State. *Encyclopedia of Philosophy*, Volume VIII New York: Collier-Macmillan: 6–11.

Biersteker, T. 1993 Evolving perspectives on international political economy: twentieth-century contexts and discontinuities. *International Political Science Review*, 14: 7–33.

Burleigh, M. 1988 *Germany Turns Eastward: A Study of Östforschung in the Third Reich*. Cambridge: Cambridge University Press.

Campbell, D. 1992 *Writing Security: United States Foreign Policy and the Politics of Identity*. Minneapolis: University of Minnesota Press.

Chabod, F. 1961 *L'idea di nazione*. Bari: Laterza.

Chomsky, N. 1992 *Deterring Democracy*. London: Verso.

Constantine, S. 1986 *Buy and Build: The Advertising Posters of the Empire Marketing Board*. London: HMSO.

Corbridge, S. E. 1986 *Capitalist World Development: A Critique of Radical Development Geography*. London: Macmillan.

Corbridge, S. E. 1993 *Debt and Development*. Oxford: Blackwell.

Cox, M. 1990 From the Truman doctrine to the second superpower détente: the rise and fall of the Cold War. *Journal of Peace Research*, 27: 25–41.

Curtis, L. 1983 *Nothing But the Same Old Story*. London: Information on Ireland.

Dalby, S. 1988 Geopolitical discourse: the Soviet Union as other. *Alternatives*, 13: 415–42.

Dallek, R. 1983 *The American Style of Foreign Policy: Cultural Politics and Foreign Affairs*. New York: Mentor.

Darwin, C. 1839 *Journal of Researches into the Geology and Natural History of the Various Countries Visited by the H.M.S. Beagle*. London: Henry Colburn.

Deibel, T. L. 1992 Strategies before containment: patterns for the future. *International Security*, 16: 79–108.

Esposito, J. L. 1992 *The Islamic Threat: Myth or Reality*. New York: Oxford University Press.

Fairgrieve, J. 1932 *Geography of World Power*. London: University of London Press.

Friedberg, A. L. 1988 *The Weary Titan: Britain and the Experience of Relative Decline, 1895–1905*. Princeton, NJ: Princeton University Press.

Friedrich, P. 1989 Language, ideology, and political economy. *American Anthropologist*, 91: 295–312.

Gaddis, J. L. 1982 *Strategies of Containment*. New York: Oxford University Press.

Gaddis, J. L. 1987 Introduction: the evolution of containment, in T. L. Deibel and J. L. Gaddis (eds) *Containing the Soviet Union*. Washington, DC: Pergamon-Brassey's.

Gervasi, T. 1988 *Soviet Military Power: The Pentagon's Propaganda Document Annotated and Corrected*. New York: Vintage.

Gilman, S. L. 1992 Plague in Germany, 1939/1989: cultural images of race, space, and disease, in A. Parker *et al.* (eds) *Nationalism and Sexualities*. London: Routledge.

Gorbachev, M. S. 1987–88 The reality and guarantees of a secure world. *Foreign Broadcast Information Service* SOV – 87 – 180, 17 September 1987.

Gregory, R. 1978 The domino theory, in A. DeConde (ed.) *Encyclopedia of American Foreign Policy*. Volume I, New York: Charles Scribner's Sons.

Grunberg, I. 1990 Exploring the 'myth' of hegemonic stability. *International Organization*, 44: 431–77.

Hall, P. (ed.) 1989 *The Political Power of Economic Ideas: Keynesianism Across Nations*. Princeton, NJ: Princeton University Press.

Hay, D. 1968 *The Age of the Renaissance*. London: Thames and Hudson.

Heidegger, M. 1959 *An Introduction to Metaphysics*. New Haven: Yale University Press.

Henrikson, A. 1991 Mental maps, in M. J. Hogan and T. G. Patterson (eds) *Explaining the History of American Foreign Relations*. Cambridge: Cambridge University Press.

Holdar, S. 1992 The ideal state and the power of geography: the life and work of Rudolf Kjellén. *Political Geography*, 11: 307–23.

Holzman, F. D. 1989 Politics and guesswork: the CIA and DIA estimates of Soviet military spending. *International Security*, 13: 101–31.

Huntington, S. P. 1993 The clash of civilizations? *Foreign Affairs*, 72: 22–49.

Kearns, G. 1993 Prologue: fin de siècle geopolitics: Mackinder, Hobson and theories of global closure, in P. J. Taylor (ed.) *The Political Geography of the Twentieth Century*. London: Belhaven Press.

Kennan, G. [Mr. X] 1947 The sources of Soviet conduct. *Foreign Affairs*, 25: 566–82.

Kern, S. 1983 *The Culture of Time and Space, 1880–1918*. Cambridge, MA: Harvard University Press.

Kristof, L. 1968 The Russian image of Russia, in C. A. Fisher (ed.) *Essays in Political Geography*. London: Methuen.

Kuhl, S. 1994 *The Nazi Connection: Eugenics, American Racism, and German National Socialism*. New York: Oxford University Press.

Luke, T. W. 1996 Governmentality and contragovernmentality: rethinking sovereignty and territoriality after the Cold War. *Political Geography*, 15: 491–507.

Mosse, G. L. 1980 *Masses and Man: Nationalist and Fascist Perceptions of Reality*. Detroit, MI: Wayne State University Press.

Norris, C. 1992 *Uncritical Theory: Postmodernism, Intellectuals, and the Gulf War*. Amherst, MA: University of Massachusetts Press.

O'Loughlin, J. and Grant, R. 1990 The political geography of presidential speeches, 1946–1987. *Annals of the Association of American Geographers*, 80: 504–30.

O Tuathail, G. 1992 Putting Mackinder in his place: material transformation and myth. *Political Geography*, 11: 100–18.

O Tuathail, G. 1993 Japan as threat: geo-economic discourses on the US–Japan relationship in US civil society, 1987–1991, in C. H. Williams (ed.) *The Political Geography of the New World Order*. London: Belhaven Press.

O Tuathail, G. and Agnew, J. A. 1992 Geopolitics and discourse: practical geopolitical reasoning in American foreign policy. *Political Geography*, 11: 190–204.

Pletsch, C. E. 1981 The Three Worlds, or the division of social scientific labor, circa 1950–1975. *Comparative Studies in Society and History*, 23: 565–90.

Pratt, J. W. 1935 The ideology of American expansion, in A. Craven (ed.) *Essays in Memory of William E. Dodd*. Chicago: University of Chicago Press.

Pratt, M. L. 1992 *Imperial Eyes: Travel Writing and Transculturation*. London: Routledge.

Renan, E. 1882 Qu'est-ce qu'une nation?, in *Discours et conferences, Œuvres complètes*, vol. I. Paris.

Rosenberg, E. S. 1982 *Spreading the American Dream: American Economic and Cultural Expansion, 1890–1945*. New York: Hill and Wang.

Said, E. W. 1978 *Orientalism*. New York: Vintage.

Sanders, J. W. 1983 *Peddlers of Crisis: The Committee on the Present Danger*. Boston: South End Press.

Schmitter, P. 1971 Still the century of corporatism? *Review of Politics*, 36: 85–131.

Sharp, J. P. 1993 Publishing American identity: popular geopolitics, myth and the *Reader's Digest*. *Political Geography*, 12: 491–503.

Sharp, J. P. 1996 Hegemony, popular culture and geopolitics: the *Reader's Digest* and the construction of danger. *Political Geography*, 15: 557–70.

Slater, J. 1987 Dominos in Central America: will they fall? Does it matter? *International Security*, 12: 105–34.

Smith, A. D. 1991 *National Identity*. London: Penguin.

Smith, G. 1994 *The Last Years of the Monroe Doctrine, 1945–1993*. New York: Hill and Wang.

Smith, W. D. 1980 Friedrich Ratzel and the origins of Lebensraum. *German Studies Review*, 3: 51–68.

Smith, W. D. 1986 *The Ideological Origins of Nazi Imperialism*. New York: Oxford University Press.

Spengler, O. 1926 *The Decline of the West*. New York: Knopf.

Spurr, D. 1993 *The Rhetoric of Empire: Colonial Discourse in Journalism, Travel Writing, and Imperial Administration*. Durham, NC: Duke University Press.

Stephanson, A. 1995 *Manifest Destiny: American Expansion and the Empire of Right*. New York: Hill and Wang.

Stokes, E. 1959 *The English Utilitarians and India*. London: Oxford University Press.

Thompson, J. A. 1992 The exaggeration of American vulnerability: the anatomy of a tradition. *Diplomatic History*, 16: 23–43.

Toulmin, S. 1990 *Cosmopolis: The Hidden Agenda of Modernity*. Chicago: University of Chicago Press.

Wills, G. 1992. *Lincoln at Gettysburg: The Words that Remade America*. New York: Simon and Schuster.

Yahil, L. 1990. *The Holocaust: The Fate of European Jewry*. New York: Oxford University Press.

INDEX

Note: Page numbers in *italic* refer to illustrations.